A Course in
Invertebrate Zoology

A GUIDE TO THE DISSECTION AND COMPARATIVE STUDY OF
INVERTEBRATE ANIMALS

A Course in Invertebrate Zoology

A GUIDE TO THE DISSECTION AND COMPARATIVE STUDY OF INVERTEBRATE ANIMALS

By

Henry Sherring Pratt, Ph.D.

Professor of Biology at Haverford College and instructor in
comparative Anatomy at the Marine Biological Laboratory
of the Brooklyn Institute of Arts and sciences at cold spring Harbor, L.I.

Revised Edition

MJP
PUBLISHERS

Chennai   Trichy   Tirunelveli   New Delhi

First published: 1910

ISBN 978-81-8094-139-9   **MJP Publishers**

All rights reserved   No. 44, Nallathambi Street,
Triplicane, Chennai 600 005

MJP 128   © Publishers, 2020

Publisher : C. Janarthanan

PREFACE

THE plan of this course is to study each of the larger groups of invertebrate animals, so far as possible, as a whole, instead of detached types of different groups taken more or less at random, as is usually done. The attention is directed constantly to the main structural features which characterize the entire group under consideration. The effort is thus made to teach relationships, and to make the study truly comparative.

In order that the systematic position of the animals examined and their larger affinities may be easily kept in mind, a synopsis of the animal kingdom expressing the relationships of the various groups has been added in an appendix.

The course begins with arthropods, because the natural succession of forms from the lowest to the highest is more apparent in them than in any other group of invertebrates, and it is, consequently, easier for a beginner, by studying them, to learn to appreciate the real significance of the blood-relationship of animals. Arthropods are also perhaps the most convenient animals with which to teach the fundamental principles of invertebrate morphology. Whether, however, the student begins his course with insects or with crustaceans, and whether the first insect taken up is the wasp or the grasshopper, will be matters for the decision of the teacher. The course has been so arranged that any of these methods of beginning may be adopted.

While the comparative feature runs through all the dissections in the course, each one is usually complete in itself and does not depend upon any others. The teacher is thus enabled to give his class such dissections as he wishes and is not compelled to adopt the entire series in order to have his course complete. In my own classes, I vary the order of the dissections from year to year and never go through the entire course. I even occasionally begin the course with the Protozoa and work upward to the higher animals; but I do not consider this usually so profitable a method of procedure for the pupil as the one herein recommended.

An important feature of the plan of this course has been adopted, in a somewhat different form, from Huxley and Martin's "Practical Biology" and Marshall and Hurst's "Practical Zoölogy." It is to give the student such practical directions that he can go on with his work intelligently and profitably without having an instructor constantly at his elbow. It has been my experience that far too much of the time of the average youthful student is often wasted in the laboratory because the instructor does not happen to be at hand at critical times to direct his work. The student will often do the work wrong in consequence, or perhaps he will not do anything at all; in either case his time is wasted and perhaps his material spoiled.

In most of the dissections the directions are so arranged that the student can complete the study with a single specimen, and the order in which the different systems of organs are taken up in each dissection is made dependent upon this feature. The necessity of practicing economy of material is thus inculcated, and the habit is acquired of studying and handling each specimen with care and judgment.

I have been fortunate in procuring the coöperation of a number of well-known teachers in the revision of the proofs, with the aid of whom I have sought to eliminate errors so far as possible. Portions of the proofs have been read critically by

Professors A. S. Packard, J. H. Comstock, H. H. Wilder, J. I. Hamaker, Frank Smith, H. B. Ward, E. L. Rice, H. L. Osborn, H. L. Clark, C. W. Hargitt, and H. S. Jennings. Their criticisms and suggestions have been most helpful and important, and I wish to acknowledge a heavy obligation to each of them.

H. S. PRATT

HAVERFORD, PA.
October, 1901

PREFACE TO THE REVISED EDITION

The principal differences between the first and second editions of this work consist in the addition of several dissections in the second edition, and the revision of the scheme of classification in the Appendix. The additional dissections are those of the house fly, a spider, the oyster, a sea cucumber, Gonionemus, and a sea anemone.

HAVERFORD, PA.
June, 1915

CONTENTS

CHAPTER I

ARTHROPODA

CHAPTER II

ANNELIDA

CHAPTER III

PLATHELMINTHES

CHAPTER IV

BRYOZOA (POLYZOA)

CHAPTER V

MOLLUSCA

CHAPTER VI

TUNICATA

CHAPTER VII

ECHINODERMATA

CHAPTER VIII

CNIDARIA

CHAPTER IX

SPONGIARIA

CHAPTER X

PROTOZOA

APPENDIX

APPARATUS AND MATERIAL

THE apparatus necessary for a course in invertebrate zoölogy need not be extensive. Each student should be provided with the following instruments : two scalpels, a small one and one of medium size ; two pairs of scissors, a large straight pair and a small pair preferably with curved tips ; two pairs of forceps, a small pair and one of medium size, both straight and with corrugated tips ; one or two dissecting needles, a probe, a blow-pipe, a hand lens.

Each student should have a shallow dissecting pan, in the bottom of which is a layer of black wax ; the depth of the pan should be about an inch and a half. If the lobster be dissected, however, a deeper pan will also be needed. The student should also be provided with a number of pins of several sizes, which may be conveniently kept, while not in use, stuck in a large cork.

It is intended that most of the drawings of dissections should be outlines, usually more or less diagrammatic, made with a hard drawing pencil in a large blank book, the paper of which is good and firm, or upon sheets of drawing paper. The general use of colors by a class is not recommended, not because the use of them is not often helpful, but because in a class of young students it is difficult to prevent their abuse by many. The careless or slothful student will often be tempted to substitute the use of colors for careful drawing. Outline drawings of a dissection on a sufficiently large scale, and carefully made and labeled, will invariably be perfectly clear.

For the study of many of the animals or parts of them in this course, a compound microscope will be needed ; a dissecting micro-scope will also be most useful throughout the course, although not indispensable. The student should be provided with a number of glass slides and thick cover-glasses. Water may be used as a medium for making temporary mounts of most of the objects examined under the microscope. A solution made of equal parts of water and glycerine, however, is usually preferable to water, as it will not dry up and, besides, renders the object more transparent. None of the animals studied here need to be stained and mounted in balsam or other permanent medium. In the case, however, of the tapeworm, the hydroids, and perhaps one or two of the other forms, the animal can be studied with greater profit if thus stained and mounted, and it is recommended that the student be provided with such specimens.

As a rule the material needed can be easily obtained. Most of the animals studied may be purchased from the supply department of the Marine Biological Laboratory at Woods Hole, Mass.; F. D. Lambert, Tufts College, Mass.; H. M. Stephens, Carlisle, Pa.; or other dealers in such supplies. Blackford's, Fulton Market, New York City, will furnish the crayfish, the lobster, the edible crab, the French snail (*Helix pomatia*), and the squid. Powers & Powers, Station A, Lincoln, Neb., will furnish live protozoans and hydras.

INVERTEBRATE ZOÖLOGY

CHAPTER I

ARTHROPODA

INSECTA

A HYMENOPTEROUS INSECT. A WASP

Observe the shape, color, and external anatomy of the animal. It is bilaterally symmetrical, *i.e.*, it has a right and a left side which are alike ; it has a dorsal and a ventral side which are unlike, and also a forward and a hinder end which are unlike, the forward or anterior end being distinguished by the possession of important organs of special sense and the mouth. All of these features are characteristic of rapidly moving animals. Can you explain why? On the ventral side are the legs, which are also called appendages or extremities. On the dorsal side of the insect are the wings, which are not called extremities, since only those organs receive this designation, speaking strictly, which are paired projections from the lateral or the ventral surface of the body, and are either used for locomotion or are homologous to locomotory organs, *i.e.*, are directly descended from organs which were primarily used for locomotion. Thus, the wings of bats and birds are extremities, although those of insects are not.

The external surface of the animal is very smooth. This feature is also correlated with rapid motion. Do you know how? The animal is encased in a hard shell, called the cuticula, which is composed largely of a very hard and resistant substance called chitin, and serves the double purpose of a protection for the internal soft parts and a surface for the attachment of muscles. It is, in fact, the skeleton of the animal, and is called an exoskeleton, in contradistinction to an internal supporting structure which would be called an endoskeleton. All invertebrate animals, except some of the lowest, are provided with a cuticular exoskeleton, but it is only the arthropods in which it is composed largely of chitin. In fact, the possession of such a hard and resistant external covering is one of the reasons why insects have so successfully maintained themselves in the universal struggle for existence.

Observe that the body of the animal is composed of a number of serially arranged segments. These are called somites or metameres, and the segmented type of structure presented by the insect body is called a metameric type of structure. Observe that the body is sharply divided into three divisions — the **head, thorax,** and **abdomen.**

The **head** is unsegmented and bears on its anterior and dorsal surface a pair of long, jointed **feelers** or **antennæ,** which are important sense-organs, a pair of large **compound eyes,** and three small, dot-like eyes, called **ocelli,** which it may be necessary to look for with a hand lens; on its ventral side are the **mouth-parts,** the organs which taste, grasp, and masticate the food. Examine these mouth-parts carefully with a hand lens; notice that there is a short overhanging **upper lip,** beneath which is a pair of powerful jaws having a lateral or side position instead of a dorso-ventral one like the jaws of vertebrates. Beneath the jaws are two other pairs of mouth-parts, the **maxillæ** and the **under lip,** which, however, will not be studied at present; notice the two pairs of elongated and segmented **palps,** which are probably organs of taste.

The **thorax** is composed of three somites or metameres, which are called, respectively, the **pro-, meso-,** and **metathorax.** Each somite bears a pair of legs on its ventral surface, and the meso- and metathorax bear each a pair of wings on the dorsal surface ; it is thus in the thorax that the organs of locomotion of the animal are concentrated. Find the sutures between the thoracic segments. The dorsal cuticula of each thoracic segment is called the **tergum** ; the ventral cuticula, the **sternum** ; and that of each lateral side, the **pleurum.** Thus we speak of the pro-, meso-, and metasternum, etc.

In the abdomen the dorsal and the ventral portions of the cuticula are composed each of a distinct plate in each somite, which are called the **tergite** and the **sternite,** respectively. The abdomen bears no appendages ; it contains most of the vegeta- tive organs of the animal. At its hinder end are the **vent** or **anus** and, in the female, the **sting.** Do you find a straight row of minute dots on each side of the abdomen and the thorax ? These are the **spiracles,** the external openings of the tracheal or respiratory system. In dark-colored wasps it may be impos- sible to see them with a hand lens, and it may be necessary to remove the cuticula from the side of the body and examine it under a compound microscope. How many are there on each side, and what relation do they bear to the segments ?

Exercise 1. Draw an outline of the side view of the wasp on a scale of 4 or 5, indicating the segmentation and all the parts observed. The three thoracic segments may be difficult to distinguish at first, but if it be kept in mind that each one of them bears a pair of legs, the task will be easy. Number on your drawing the thoracic and abdominal segments, and carefully label all the different parts and organs.

Exercise 2. Draw an outline of the face on a scale of 10, showing exactly the relative length and the segmentation of the antennæ, the position of the compound eyes and ocelli and the upper lip, and label them all.

Exercise 3. Remove a metathoracic leg and draw an outline
of it on a scale of 5. Its different segments, beginning
with the proximal one, *i.e.*, the one nearest the body, are
the following : the **coxa**, by which the leg articulates with
the body ; the **trochanter**, a very small segment ; the **femur**
or thigh, a long segment ; the **tibia** or shank, also long ; the
tarsus or foot, which is composed of five small segments,
the last one of which bears the two claws. Label all of
these.

Exercise 4. Remove a mesothoracic wing, extend it, and draw a
picture of it on a scale of 5, indicating its venation.

Save your specimen in a dish of formalin or alcohol for future
use. We shall reserve the detailed study of the mouth-parts
until the grasshopper is taken up, when the mouth-parts of the
various orders of insects will be studied together. The internal
anatomy of all insects is exceedingly similar, and it will not be
necessary to study it in more than one animal; we select the
grasshopper as being the one best suited.

INSECTA

A COLEOPTEROUS INSECT. A LARGE BEETLE

Compare the animal with the wasp. We notice, in the first place, the heavier and clumsier body and the smaller head. The animal is evidently much less active and also less intelligent than the wasp. We notice, also, that the wings lie close to the body instead of being raised above it. The forward or mesothoracic wings are hard and thick; they are not used for flight, but cover the metathoracic pair and the hinder part of the body and are called the **wing-covers** or **elytra**. They form, thus, an additional protection to the back. The entire body of most beetles, in fact, has a thicker cuticula and, consequently, a more effective external covering than that of the wasp. This feature may be correlated with the smaller intelligence of the animal. Opening the elytra, we notice beneath them the membranous metathoracic wings with which the animal flies; we notice also that they are folded transversely as well as longitudinally. These wings are wanting in some of the running beetles, where the wing-covers are sometimes fused. Note the **scutellum**, the small triangular plate, between the base of the wing-covers. Find the **eyes** and note their small size. Are **ocelli** present? Find the **antennæ**; in some beetles they are often concealed beneath the sides of the head.

Exercise 1. Draw an outline of the dorsal aspect of the beetle on a scale of 4 or 5. First, however, spread and pin the right wing-cover and wing. Number the thoracic and abdominal segments and label all the parts observed.

Exercise 2. Draw an outline on the same scale of the ventral aspect of your beetle, tracing carefully the sutures between the segments. Number the thoracic and abdominal segments.

Exercise 3. Remove a mesothoracic leg and draw an outline of it on the scale of 5. Label the segments.

Exercise 4. Remove a wing and draw an outline of it on a scale of 5, tracing in the veins.

Save your specimen in formalin or alcohol for future use.

INSECTA

A DIPTEROUS INSECT. THE FLY

Kill several bluebottle flies or large house flies, without injuring them, and impale one on a slender insect pin or a needle. Stick the pin or needle into a cork or a small piece of wood, in order to be able to handle it easily, and study the external anatomy of the fly with the aid of a hand lens.

Observe the compact body of the animal, and note that it is distinctly divided, like that of the wasp, into three divisions — the **head,** the **thorax,** and the **abdomen.** Observe the color and the hairy surface of the body, including the legs and the wings. These numerous **hairs** are projections of the cuticula, and perform a useful function as tactile organs; that is, they are sensitive to vibrations of the atmosphere, and thus function as sense organs in that they aid in giving the animal a knowledge of its surroundings. Note the three pairs of long, strong **legs** and the single pair of **wings.** The fly has unusual locomotory powers. Correlated with these powers are the long cuticular hairs just mentioned, and also the very large composite **eyes.** An active, rapidly moving animal like the fly needs well-developed organs of orientation. The eyes are larger in the male than in the female, and are closer together on the top of the head. The two sexes may thus be distinguished.

Between the large eyes are the three minute accessory eyes or **ocelli.** Note the peculiar form of the small **antennæ,** with their pinnate terminal portion. Extend the **proboscis** and observe its complex structure and the **oral lobes** at the lower end. The fly eats only fluid food, which it sucks up through its proboscis.

The **thorax** is of relatively large size, being almost entirely filled with the very extensive musculature of the legs and wings. The three thoracic somites are of unequal size. The middle one is the largest and bears the wings. Note that the hinder margin of the basal portion of the wing is divided into three prominent lobes. The posterior thoracic somite is the smallest and bears the **balancers**, which are the morphological equivalents of the second pair of wings, possessed by most insects. These are a pair of minute white, knobbed organs, which project backward from the posterior wall of the somite, each one being covered by the basal lobe of the wing on that side. They have a sensory function.

The **abdomen** is composed of eight somites in the male fly and nine in the female. Of these, however, four somites are much larger than the others, and make up the greater part of the abdomen. The sixth, seventh, and eighth in the male are very small and rudimentary. In the female the posterior four form a long, tubular **ovipositor**, which is usually telescoped into the abdomen but can often be squeezed out by a little pressure. Each of the five anterior abdominal somites has a pair of **spiracles**. Find them.

Exercise 1. Draw an outline of the dorsal aspect of the fly on a scale of about 10, indicating the segmentation and the parts observed, including the venation of the wings. Label all the parts observed.

Exercise 2. Turn the fly over on its back and draw one of its legs on a large scale. The names of the different segments of the leg may be obtained from Exercise 3 on page 4. Note, between the two claws on each foot, the two **pulvilli** — the hairy adhesive pads by means of whose sticky secretions the fly can walk on an inverted surface.

Exercise 3. Draw, on a large scale, a side view of the head with the proboscis extended. Note carefully the form of the antennæ and of the proboscis. The latter is homologous to the under lip or labium of other insects.

INSECTA

AN ORTHOPTEROUS INSECT. A LARGE GRASSHOPPER

Observe the shape, color, and external anatomy of the animal. Note the long, vermiform body and the large head. The body, as in all insects, is made up of a number of serially arranged segments, called **somites** or **metameres**, which fall into two divisions — the thorax and the abdomen. The **head** is unsegmented, being composed of a number of completely fused somites, and bears upon its dorsal and anterior surface a pair of long, jointed feelers or **antennæ**, which are important sense-organs, a pair of large compound **eyes**, and three small, dot-like eyes, called **ocelli**, which it may be necessary to look for with a hand lens; on its ventral side are the **mouth-parts**, the organs with which it tastes, grasps, and masticates its food. Examine these mouth-parts with a hand lens. Observe the long, broad **upper lip** and pass a needle under its ventral edge. Back of the upper lip will be seen the strong **mandibles**, and by pressing these to the right and left the two remaining pairs of mouth-parts, the **maxillæ** and the **under lip**, will be seen. Note the two pairs of jointed **palps** belonging to them, which are probably organs of taste. These parts will all be studied later in detail.

The **thorax** is made up of three somites, which are called the **pro-, meso-,** and **metathorax.** Notice that the thorax is not separated from the abdomen by a constriction, as it is in the wasp, but, however, that it may be easily distinguished from the abdomen by its greater diameter. The prothorax is movable, as in the beetle, and its dorsal and lateral surfaces are covered by a large shield. On the ventral side of the prothorax, between the prothoracic legs, is, in many grasshoppers, a short projection. The meso- and metathorax are united immovably with the abdomen and are covered by the two pairs of wings. The anterior or mesothoracic wings are parchment-like and are not functional in flying, but, like the wing-covers of beetles, are held out at right angles to the body during flight. The metathoracic wings are membranous and are folded longitudinally like a fan beneath the forward wings, when at rest. Each somite bears a pair of legs on its ventral surface. The cuticula of each thoracic somite is composed of a number of distinct plates. Those which constitute the dorsal and the ventral surfaces form the **tergum** and the **sternum** of the somite, respectively; those constituting the lateral surfaces form the **pleura** of the somite. Thus we speak of the pro-, meso-, and metasternum, etc.

In the abdomen the cuticula of the dorsal and the ventral portions of each somite is composed of a single plate, which is called the **tergite** and the **sternite**, respectively. The **abdomen** is made up of eleven somites, which are not all, however, perfect segments, the sternite of several of the terminal somites being wanting. The posterior end of the abdomen is different in the two sexes, the female possessing an ovipositor, by means of which she buries her eggs in the ground. The sternites of the ninth, tenth, and eleventh somites are wanting in the female, the last sternite being the eighth. Tergites of the three terminal somites are, however, present. Projecting from the hinder end of the abdomen is the **ovipositor**, which consists of two pairs of short, movable, curved, and pointed structures. One of these pairs is dorsal in position, and the **anus** is at its base; the other is ventral, and at its base is the external opening of the **oviduct**. Extending from the posterior border of the tenth tergite is another pair of pointed projections, called **cerci**, which may have a sensory function. Just beneath each cercus is a plate called a **podical plate**. Between the two podical plates on the dorsal side of the animal is the triangular eleventh tergite.

In the male the ninth and tenth sternites are present, although they may be fused so as to appear as one plate. An additional ventral plate, called the **genital plate**, forms the posterior extremity of the body. The tenth tergite is very small; the **podical plates** and the **cerci** are large. Beneath the eleventh tergite is the **anus**. Compare a male with a female abdomen and identify the parts above mentioned.

On the lateral side of the first abdominal segment note the **auditory organ**, a large circular opening covered by a membrane. With the aid of a hand lens find the **spiracles** of the thorax and the abdomen. Ten pairs are present, one pair on the anterior margin of both the meso- and the metathoracic segments, and one pair on each of the eight anterior abdominal segments, that on the first abdominal segment being just within the margin of the auditory organ.

Exercise 1. Spread out and pin down all four wings and draw an outline of the dorsal aspect of the grasshopper on a scale of 2 to 4. Number the thoracic and the abdominal segments, and label all the parts observed.

Exercise 2. Cut off the wings from the left side of the body and draw an outline of the side view of the thorax and the two anterior abdominal segments on a scale of 5 or 6. Note that both the meso- and the metapleurum are divided by a diagonal suture into two portions. Number the segments and label all the parts.

Exercise 3. Draw a side view of the posterior end of your specimen (whether male or female) on a scale of 5 or 6, showing accurately the arrangement of all the parts, and label them all.

Exercise 4. Draw an outline of the ventral surface of the thorax on a scale of 5 or 6. Note the dovetailing of the anterior margin of the metasternum with the posterior margin of the mesosternum and of that of the first sternite with the metasternum, also the attachment of the legs.

Exercise 5. Remove a metathoracic leg and draw an outline of it on a scale of 3. The segment by which it articulates with the body is the **coxa**; the next segment is the **trochanter**, which in the grasshopper, however, is not a free segment, but is fused with the following one, the **femur**; the latter is the largest segment of the leg and has V-shaped muscle impressions on its surface; the next segment is the **tibia** or shank; the end segment is the **tarsus** or foot, which is made up of five smaller segments; the terminal one of these bears two claws between which is a structure called the **pulvillus**. This organ is an adhesive pad which enables the animal to walk and spring on smooth surfaces. Label all of these parts.

Exercise 6. Draw an outline of the face on a scale of 5 or 6. The large plate which forms the top, front, and sides of the head, in which the eyes, ocelli, and antennæ are situated, is called the **epicranium**. The sides of the epicranium, back of the eyes, are the genæ, the top is the **vertex**, and that part which forms the anterior surface is the **front**. Ventral to the epicranium is a broad, short, median plate called the **clypeus**, beneath which is the **upper lip**. The antennæ are the first pair of appendages. Label all parts.

The **mouth-parts**. These consist of the median upper lip or **labrum**, the paired **mandibles**, the paired **maxillæ**, the median **hypopharynx**, and the paired under lip or **labium**. The paired mouth-parts are the second, third, and fourth pairs of appendages, the antennæ being the first pair.

Exercise 7. Remove the labrum with scissors and draw it on a scale of 5.

Exercise 8. With strong forceps remove the dark-colored mandibles and draw the inner surface of one of them on a scale of 5.

Exercise 9. Remove the maxillæ, which lie just back of the mandibles, being careful to take out the entire structure. Mount them on a glass slide in glycerine or water with the posterior side uppermost, and examine them under the microscope. Note the following parts: the basal segment or **cardo**, by which the maxilla articulates with the head; the **stipes**, the broadest segment of the structure; the **inner** and the **outer lobes**, which project from the distal edge of the stipes; and the **maxillary palp**, which projects from the lateral edge of the stipes. Draw a maxilla on a scale of 5 and label all of these parts.

Exercise 10. Note between the maxillæ and just in front of the labium a median projection, the hypopharynx. Remove the labium, taking care to leave none of it in the animal, mount it on a slide, and identify the following parts: the basal segment or **submentum**, by means of which the labium articulates with the head; the **mentum**, the succeeding segment; the **ligula**, which projects from the distal edge of the mentum; and the two **labial palps**, which project from the lateral edges of the mentum. The labium is a second pair of maxillæ fused in the median line. Trace the homologies between the parts of the labium and those of the maxillæ. Draw the labium on a scale of 5 and label its parts.

The mouth-parts of the wasp and the beetle. The mouth-parts of the grasshopper are called biting mouth-parts because the insect bites or chews its food instead of licking or sucking it. Biting mouth-parts characterize all the more primitive insects. The mouth-parts of the beetle are similar to those of the grasshopper, although the former is a much higher insect.

Exercise 11. Remove carefully and with the aid of the dissecting microscope, if necessary, the antennæ, labrum, mandibles, maxillæ, and labium of the beetle. Mount them on a slide and draw them on a large scale. Label carefully all the parts.

Exercise 12. The mouth-parts of the wasp are much more highly specialized than those of the beetle, as they are adapted not only for chewing, but also for licking. Remove the antennæ and the mouth-parts of the wasp and mount them on a slide. The labrum and the mandibles will be seen to be similar to those already studied. The maxillæ and the labium, also, do not differ materially from those of the beetle or the grasshopper. The labium lies between the two maxillæ, and its ligula is elongated and modified to form a licking organ. Draw an antenna and the mouth-parts on a scale of 6.

Internal anatomy. Take the grasshopper in the hand and with a pair of fine, sharp scissors cut a slit through the body-wall a little to one side of the mid-dorsal line from one end of the body to the other, using great care not to injure the organs within. Place the animal, dorsal side up, in a shallow pan with a wax-covered bottom containing water or 30% alcohol. First, with two strong pins, pin the head to the wax and then the extreme hinder end of the body, then carefully spread the cut edges of the body-wall as widely as possible to the right and left and pin them down, using many pins on each side. Observe the organs as they lie in the body-cavity. In the thorax will be seen the strong locomotory muscles. Lying immediately beneath the dorsal abdominal wall in the median line is the **heart**; this may have been destroyed by the incision, but if not, it may be recognized as a narrow, transparent tube of the diameter of a needle, flanked by paired triangular muscles

which hold it to the body-wall.   Immediately beneath the heart
is a loose network of yellowish fatty tissue, called the **fat-body,**
which covers the viscera.   Remove this carefully.   The **alimentary canal** will be disclosed, a large tube running through the
median axis of the body; above the abdominal portion are the
paired **reproductive glands,** from which a **duct** passes on each
side around the alimentary canal to the ventral side of the
animal.   Notice the silvery air-tubes or **tracheæ** and the **air-sacs**
on each side of the alimentary canal; also observe the tangled
mass of delicate brown threads, the urinary or **Malpighian
tubules,** between the reproductive glands and the alimentary
canal.

Exercise 13. Make a sketch of the animal on a scale of 5, showing the internal organs *in situ,* and label them all.

The digestive system.   With fine scissors sever the alimentary
canal at its extreme posterior end, where it joins the anus.
With great care draw it forward between the ducts of the
reproductive organs and from beneath those organs, loosening
it from the surrounding tissues with a needle.   Identify the
following divisions of the alimentary canal: the **pharynx,** the
space just back of the mouth; the **œsophagus,** the narrow tube
which runs upward from the pharynx and, bending back,
enters the thorax, where it enlarges to form a pouch called the
crop; the **salivary glands,** a pair of delicate, branched organs, one
on each side of the crop, the **ducts** of which run forward to the
pharynx; the **gastric cœca,** eight elongated sacs which encircle
the base of the crop; the **stomach-intestine,** a large tube which
extends back to the point where the delicate urinary or
Malpighian tubules join the alimentary canal; the **ileum,** a
thick tube the diameter of which is the same as that of the
stomach; the **colon,** a narrow, slightly coiled tube; and the
rectum, which has six ridge-like **rectal glands** along its sides and
opens into the **anus.**

The excretory system. This system consists of the **Malpighian tubules**. These are delicate tubular glands, about fifty in number, which unite with and discharge their products into the alimentary canal at the point of juncture of the stomach-intestine and the ileum. They extend freely into the body-cavity and excrete urinary wastes from the blood, in which they lie immersed.

Exercise 14. Make a drawing of the alimentary canal and the Malpighian tubules on a scale of 7 and label all of the parts.

The reproductive system; the female organs. The two **ovaries** are closely bound together by a web of connective tissue and tracheæ so as to form a single mass, which lies above the intestine. If your specimen be a female, part this mass along the median line and with a needle gently remove some of the connective tissue surrounding it. Examine it with a hand lens; each side is a separate ovary and will be seen to be a collection of parallel, tapering tubules, their smaller ends being in the median line, their longer ends projecting back to the tube-like oviduct. These tubules are called **ovarioles**; it is in them that the eggs develop. How many tubules do you count on each side? Notice the elongated eggs in each ovariole. How many do you see in each one? The two **oviducts** proceed from the ovaries to the ventral side of the animal, where they unite to form a median tube, the **vagina**, which opens to the outside between the **ovipositors**. Just above the vagina is a small sac, the **receptaculum seminis**, which is connected by a long sinuous duct with the exterior. This sac becomes filled with spermatozoa during pairing, which fertilize the eggs as they pass out of the vagina.

Exercise 15. (*a*) Make a semidiagrammatic drawing representing all the parts of the female reproductive tract.

The male organs. The paired **testes** which secrete the spermatozoa lie above the intestine, bound together by connective tissue and fat. Each testis consists of a bundle of elongated tubes with which a duct called the **vas deferens** connects posteriorly. The two vasa deferentia run, one on each side of the intestine, to the ventral side of the animal, where they meet to form a median tube, called the **ductus ejaculatorius,** which is homologous to the vagina of the female. Connecting with the ductus ejaculatorius are a number of tubular prostate glands which secrete the fluid in which the spermatozoa are suspended.

Exercise 15. (b) Make a semidiagrammatic drawing representing all the parts of the male reproductive tract.

The respiratory system. The **spiracles** have already been noted. They are the external openings of the **tracheæ,** a system of fine air-tubes which extend throughout the entire body of the insect and through which fresh air is introduced into every part of the body. The blood is thus constantly aërated, and there is never any venous blood present. This arrangement results in a very active metabolism, and is one of the causes of the extraordinary amount of energy which characterizes most insects. With the aid of a hand lens examine the tracheæ in different parts of the body. They may be easily detected by their silvery gleam. Notice the arrangement of the main **tracheal trunks,** including those which connect with the spiracles, also the arrangement of the **air-sacs,** which are expansions of tracheæ. Mount a small portion of the fatty tissue containing tracheæ in water or glycerine and examine them with a compound microscope. Notice the **spiral threads** which line the tracheæ.

Exercise 16. Make a drawing of a trachea seen under a high power of the microscope.

The circulatory system. This system is very simple in insects, owing to the great complexity of the respiratory system. Instead of the blood being carried to the respiratory organs to be aërated, as is the case in vertebrates, rendering necessary a complicated system of blood-tubes connecting the remotest parts of the body with the respiratory organs, the respiratory organs are themselves a system of tubes which introduce air to every part of the body. The insect has a blood fluid which lies in the body-cavity. The only circulatory vessel present is the tubular **heart.** This organ, whose position has already been noted, has a closed hinder end and segmental **valvular openings** along its sides. By its contractions the blood is sent into the forward portions of the body, whence it flows back into the hinder portions, and enters the heart again through the valvular openings. To observe the heart of an insect is not always easy, owing to its position so near the dorsal body-wall and its great delicacy of structure. An easy method is to mount a live, transparent, aquatic insect larva, such as that of the mosquito, on a slide in water and observe it under a compound microscope. The heart and its action may be easily studied.

The nervous system. Cut off the alimentary tract at its forward end, taking care not to injure the two nerve connectives which pass to the brain, and remove all the viscera from the body. The **nerve cord** will be seen lying on the ventral body-wall of the abdomen, in the median line, slightly concealed by fat. It will be seen to be double and to contain, in the abdomen, five enlargements, the **ganglia,** from each of which fine **nerves** radiate. Trace the nerve cord from the abdomen into the thorax. It is here protected by hard projections of the body-wall, which must be carefully removed. Four large ganglia will be found here, the three posterior ones of which are the **thoracic ganglia.** The one in the forward portion of the prothorax really belongs to the head and is called the **subœsophageal ganglion.** From it a pair of nerve connectives passes to the dorsally situated **supracœsophageal ganglion** or brain. The brain is the largest ganglionic mass in the body and is situated in the top of the head between the eyes. Lay bare the brain. Notice the **optic lobes** going to the eyes, and between them the much smaller **ocellar lobes** sending nerves to the lateral ocelli. Beneath the optic lobes are the **antennal lobes,** which send nerves to the antennæ, while near them in the median line is the **median ocellar lobe,** which sends a nerve to the median ocellus.

Exercise 17. Make a large sketch of the nervous system, representing it in an outline of the animal's body, and show in which segments the different ganglia occur.

Exercise 18. Draw a diagram representing a side view of a grasshopper on a scale of 3 or 4, in which the segmentation, the relative position of the heart, the alimentary tract, and the nervous system are accurately indicated.

INSECTA

AN INSECT LARVA. A CATERPILLAR

Notice that the head, thorax, and abdomen are not set off from one another. The body is thus worm-like in form, there being almost no specialization of the body-parts. Determine how much of the body is thorax and how much abdomen. The thorax bears three pairs of jointed legs, each one terminating in a single hook. The abdomen also bears several pairs of legs which are not like those of the thorax. How many are there and in what do they differ from the thoracic legs? Find and count the spiracles, which are usually easily seen.

Exercise 1. Draw an outline representing a side view of the animal on a scale of from 2 to 6 ; number the thoracic and abdominal segments, show the spiracles, and label all the parts.

Study the **head** with the aid of a hand lens. Notice the pair of large **convex plates** which, with the small median **triangular plate**, form the wall of the head. Near the lower edge of each of the convex plates are several minute **ocelli** ; count them. On the ventral side of the head find the **antennæ**; how many joints are there in each? The mouth-parts are between the antennæ. The **labrum** is bilobed, and beneath it are the dark-colored **mandibles**. Just back of these are the **maxillæ** and the **labium**, the latter being a median, elongated, conical organ between the maxillæ. The external opening of the silk glands is in the labium.

Exercise 2. Draw a front view of the head on a scale of 7.

Internal anatomy. With fine scissors make a longitudinal incision the length of the animal, in the dorsal integument, a short distance to one side of the median line. Turn the integument to the right and left and pin it down. If it has not been destroyed, observe the **heart**. It is a straight, transparent tube lying in the mid-dorsal line just beneath the integument. Note the large, tubular **alimentary tract** surrounded by delicate, glistening tracheæ and by the white and often filamentous fat. Its forward portion is the **œsophagus**; the middle and largest portion is the **stomach-intestine**; the narrow portion back of which is the **intestine**; while the dilated portion which communicates with the anus is the **rectum**. In the forward portion of the body cavity, along the wall of the œsophagus, is a pair of delicate tubular **salivary glands** which extend forward and communicate with the mouth. Note and trace the course of the much larger tubular **silk glands** on the ventral body-wall; they are also a single pair and communicate with an opening in the labium.

Find and carefully trace the course of the six **Malpighian tubes,** which lie along the stomach and join it at its posterior end.

Exercise 3. Draw an outline of the opened animal on a scale of 6, showing the organs above described. Represent the segmentation and show accurately the position of the organs in their proper segments.

Sever the œsophagus and remove the stomach and the intestine from the body. Study the nervous system. Note the arrangement of the tracheæ with reference to the spiracles. Note the longitudinal muscle bands which form a part of the body-wall; also their segmental arrangement.

Exercise 4. Draw an outline of the opened body on a scale of 6, showing and numbering the segments. Draw in it the nervous system, representing accurately the number of ganglia, and placing them in the proper segments, together with the tracheæ and muscles.

The **reproductive system** consists of two small sexual glands and a duct leading from each. There is no external pore.

MYRIAPODA

A CHILOPOD.   A CENTIPED (*Lithobius*)

Myriapods are worm-like animals which live under logs and
stones, beneath the bark of decaying stumps and trees, and
in other dark, damp places. The two main groups of myriapods may be easily recognized by the differences in shape and
habits, — the Chilopoda being flattened and very active animals
with one pair of legs to a segment, the Diplopoda being usually
cylindrical animals with short legs, two pairs of which are
present on most of the segments.

Observe the vermiform body, the well-marked segmentation,
and the segmented legs ; also the lack of specialization among
the segments, there being no division into thorax and abdomen.
The animal is plainly an arthropod, but it is not an insect ; it
is a lower animal than an insect, because its body shows less
specialization. Note the single pair of **antennæ** and the insect-
like **mouth-parts**, also the large **hook-like appendages** just back of
the head. These latter are homologous to the first pair of
legs ; they are the principal organs of prehension and are
provided with poison glands which open on the inner surface
near the end. Note the **anal feelers** ; these are homologous to
the hindermost legs and enable the animal to perceive what is
back of it.

Exercise 1. Draw an outline of the dorsal aspect of the animal
on a scale of 5 and label all the organs observed.

Exercise 2. Draw a ventral view of the head on a scale of
10, showing the cephalic appendages in position. The
mouth-parts consist of **a** pair of **mandibles** and two pairs

of **maxillæ**, the second pair of which is homologous to the labium of insects.

Exercise 3. Remove, under a dissecting microscope, the prehensile hooks and the mouth-parts, beginning with the posterior ones and working forward, and the antennæ. Mount them on a slide and draw an outline of each. Compare the different structures of the mouth-parts with those of the insect and label them all.

The internal organs. The digestive, circulatory, respiratory, excretory, and nervous systems are essentially like the same systems in insects. The reproductive system consists of a pair of sexual glands with paired ducts, the posterior portions of which unite to form a common duct. This opens to the outside in the genital segment, which is the penultimate body segment.

ARACHNIDA

A SPIDER

As large a spider as possible should be obtained for this study. If a small one is used, it is usually well to stick a slender insect pin through it, in order to be able to handle it easily, and it should be studied with the aid of a hand lens. Observe the form and color of the animal. The body is unsegmented (although the body of the embryo spider is distinctly segmented) and is made up of two parts, the **cephalothorax** and the **abdomen**. What does the embryonic segmentation indicate as to the ultimate relationships of spiders? Observe the **hairs** which cover the body and legs. They are projections of the cuticula and are important sense organs, being sensitive to vibrations of the atmosphere. They thus aid in giving the animal information in regard to what is going on about it.

The cephalothorax. This division of the body is equivalent to the head and thorax of insects. Observe carefully the eight eyes at or near its forward end, both the size and arrangement of which vary much in the various species of spiders. The ventral surface bears the six pairs of appendages, the **mandibles,** the **pedipalps,** and the four pairs of **legs.**

The mandibles, the anterior pair, occupy a vertical position at the front end of the body and consist each of a basal portion and a terminal claw, near the tip of which is the pore from which poison is injected into the bite. In consequence of the vertical position of its mandibles the spider can only strike an insect which is beneath it.

The second pair of appendages are the **pedipalps.** These are leg-like and contain one less segment than the legs, the missing segment being the one next to the last. The basal segment of the pedipalp is called the **maxilla**. The two maxillæ are flattened structures situated on the underside of the cephalothorax just back of the mandibles, their forward, medial margins, which cover the mouth, being used to lacerate and squeeze the food so that the animal juices can be sucked up. Spiders prey exclusively upon living animals; but they can take in only liquid food. The pedipalps of the female spider differ in shape from those of the male, and the two sexes may be distinguished in this way. In the female the pedipalp looks exactly like a small leg; in the male the terminal portion is expanded and very complex in structure, being used by the animal in the act of pairing.

The third, fourth, fifth, and sixth pairs of appendages are the **legs**, each of which is composed of the following seven segments: the coxa, trochanter, femur, patella, tibia, metatarsus, and tarsus. The legs are used by the spider for a variety of purposes besides walking. They are important as tactile organs, their great length increasing their usefulness in this respect, and they undoubtedly compensate the animal in a certain degree for the lack of antennæ. They are also of use in spinning and manipulating the web, the complex structure of the claws being associated with this function.

The median plate between the maxillæ on the ventral side of the body is the **labium**; the one between the bases of the legs is the **sternum**.

The abdomen. The dorsal surface is usually marked by several pairs of depressions which mark the points of attachment of muscles. At the hinder end, on the ventral surface, are three pairs of **spinnerets**. Study these carefully with the aid of a hand lens. At the end of each spinneret are numerous microscopic holes, from which is exuded the semifluid silk. This is made up of many soft strands, which harden as they unite to form the thread.

A study of the embryology of the spider shows that the spinnerets are homologous to abdominal legs.

Note the **spiracles**, the external openings of the respiratory organs, the tracheæ and the lungs. A short distance in front of the spinnerets in the ventral surface of the abdomen is the single median, minute **tracheal spiracle**: it is often difficult to see. The **lung spiracles** are a pair of large slits near the anterior end of the abdomen, each one at the lateral end of a transverse fold of the integument. Between them in the median line is the **genital pore**. In the female spider it is covered by a large and complex plate called the **epigynum**.

Exercise 1. Cut off the legs on the right side of the body and draw an outline of a side view of the spider on a scale of from 5 to 10, putting in only the basal portion of the legs but all of the pedipalps and the mandibles. Carefully label all the parts observed.

Exercise 2. Draw an outline of the ventral aspect of the body on the same scale, putting in and labeling all the parts observed.

Exercise 3. Draw the front end of the body on a scale of 10, showing the mandibles and the eight eyes.

Exercise 4. Draw the pedipalp on a scale of 6.

Exercise 5. Draw one of the legs on a scale of 6.

Exercise 6. Cut off a tarsus and study it under a compound microscope, noting the shape of the claws and the hairs which often surround them. Draw them.

The **internal anatomy** of the spider will not be studied in this dissection. The **heart** is an elongated tube which lies, enclosed in a pericardial space, in the dorsal portion of the abdomen. From its anterior end an aorta extends into the cephalothorax and sends off a number of large branches to the legs and other organs.

The **digestive system** consists of a straight alimentary tube and its many branches. In the cephalothorax, branches of it extend to the legs, and a portion of it forms a sucking stomach, by means of which the spider sucks up its fluid food. In the abdomen it becomes the intestine and gives off an extensive network of tubules, which fills a large part of the abdomen and has the appearance of a compact gland; its function, however, is not secretory and it does not differ in structure from the rest of the intestine. The end intestine possesses a large dorsal fæcal reservoir.

The **kidneys** are a pair of branching Malpighian tubules.

The **brain** lies just beneath the eyes. It is joined, by means of broad connectives, with the large ventral ganglionic mass, from which nerves extend into the abdomen and the appendages.

The **organs of respiration** are the lungs and the tracheæ. The lungs are a pair of sacs which open to the outside through the lung spiracles, each sac containing a series of lamellæ, usually called a lung-book, in which the blood circulates. The tracheæ open to the outside through the tracheal spiracle.

The sexes are separate in spiders. In the male the **testes** are a pair of tubular glands which are joined, by means of the coiled sperm ducts, with the sperm vesicle, which opens to the outside through the genital pore. The **ovaries**, in the female, are large organs in the ventral portion of the abdomen, which are joined, by means of the oviducts, with the uterus, which, after receiving the paired receptacula seminis, opens to the outside through the genital pore.

The **silk glands** are branched or tubular structures in the ventral portion of the abdomen.

CRUSTACEA

A MACRURAN DECAPOD.   A CRAYFISH OR A LOBSTER

These two animals are very common, the one in fresh and the other in salt water. In external form and internal anatomy they are exceedingly similar to each other, and the same directions for dissection may be made to apply to either. In habits and general method of life the animals also resemble each other; they move about at or near the bottom of the water, preferring regions which are rocky or stony, and feed upon small animals of all kinds and upon carrion.

Observe the shape, color, and external anatomy of the animal. It is bilaterally symmetrical; the body is composed of a number of serially arranged segments, which are called **somites** or **metameres**; the dorsal and the ventral sides of the body are unlike, the latter being characterized by the possession of a series of paired and jointed appendages metamerically arranged; *i.e.*, each somite or metamere bears a pair of appendages; the anterior and the posterior ends are also unlike, the former being characterized by the possession of organs of special sense and the mouth. The external covering of the body is a chitinous cuticula which constitutes an exoskeleton. All of these features are equally characteristic of insects and myriapods.

As in all crustaceans, and also in insects, the body of the animal falls into three distinct divisions, — the **head, thorax,** and **abdomen.** The first two of these body-divisions do not, however, articulate freely with each other as they do in insects, but, in common with all the higher crustaceans, they are fused together and form a single structure, which is called the **cephalothorax.** The dorsal and the lateral surfaces of this division show no segmentation, because of the fusion of the somites and the presence of a hard, shield-like structure covering it, which is called the **carapace**, but on the ventral side the segmentation is distinctly seen. Extending along the entire ventral surface of the animal are the paired appendages. Their metameric significance may not be seen in the cephalothorax, but it will be distinctly seen in the abdomen, where each somite except the last bears a pair of appendages.

The cuticular exoskeleton is thicker and heavier than in insects; this is due to the presence, besides chitin, of salts of lime. The crayfish or lobster moults its cuticula periodically, the adult animal probably once or twice a year, the young animals oftener.

The animal is capable of two sorts of locomotion. By powerful strokes of the broad, fin-like end of the abdomen it swims rapidly backward, and it can walk on its thoracic legs. It is well provided with special sense-organs. Most important to it are the two pairs of feelers or **antennæ**, which are characteristic of all crustaceans, and the **compound eyes** on movable stalks. It also possesses, in a pair of small cavities, on the upper surface of the basal joints of the first or shorter pair of antennæ, peculiar sense-organs, which were formerly supposed to be ears, but are now known to be **balancing organs**. With the aid of them the animal maintains its equilibrium.

The body of the crayfish or the lobster, as of all the higher crustaceans, is made up of twenty somites or body-segments, of which the thirteen anterior somites form the cephalothorax, and the seven posterior ones the abdomen.

The cephalothorax. The anterior five somites forming this body-division are cephalic, the remaining eight are thoracic, and all are covered dorsally and laterally by the **carapace**. The projection running forward from the anterior end of the carapace is called the **rostrum**. A transverse groove is seen near its middle; this is the **cervical suture** and marks the

boundary between the head and the thorax.   In the crayfish
two semicircular, longitudinal grooves extend backward from
the outer ends of the cervical suture, which separate the sides
of the carapace from the median, dorsal portion.   The sides of
the carapace are called the **branchiostegites**; they cover lateral
folds of the dorsal integument of the animal, which extend
over the sides of the body and enclose between themselves and
it the spaces within which lie the **gills**.   These spaces, the **gill-
chambers**, thus communicate freely with the surrounding water.
Pass the handle of a scalpel or other flat object beneath the
lower edge of the branchiostegite and it will go into the gill-
chamber.   During life a current of water passes constantly
into the gill-chamber along this lower edge, where it bathes the
gills and then passes out at the forward end.

Study the ventral side of the cephalothorax.   The most
important organs here are the appendages.   At the anterior
end of the body are the two pairs of **antennæ**, the longer pair
being the second.   On the lower surface of the basal joint of
each of the latter is an opening ; these are the external openings of the kidneys or **green glands**.   Back of the antennæ is
the **mouth**.   It is bounded in front by a lip-like structure called
the **labrum**, at the sides by the strong **mandibles**, and behind by
a pair of delicate plate-like projections, called the **paragnatha**,
which are not appendages.   Press the mandibles aside and
pass a probe into the mouth.   Between the mouth and the
large claws are five pairs of appendages which assist in the
act of eating ; they are two pairs of delicate leaf-like **maxillæ**,
just back of the mouth, and three pairs of larger **maxillipeds**,
back of them.   They are best identified by beginning with
the hinder pair of maxillipeds, which is just in front of the
large claws, and working forward, placing a needle or knife
between the appendages as they are identified.   Back of the
maxillipeds come the large grasping claws or **chelipeds**, which form
the principal weapons of offense and defense of the animal, and

in the largest lobsters are powerful enough to crush a man's arm. Note the difference between the right and the left claw, if any. Back of the chelipeds are four pairs of walking legs. In the male animal the paired external openings of the genital organs are at the base of the last pair of walking legs, in the female at the base of the antepenultimate pair. Find them.

The abdomen. The seven somites forming this body-division are all free and jointed with one another. Note the difference in the thickness of the cuticula on the dorsal and the ventral surfaces, also its thinness at the joints. The appendages on the abdomen have various uses. They probably have a general respiratory function. In the male the first two pairs are functional in pairing, in the female the first five pairs hold the eggs from the time they are laid until the young are hatched. The last pair in both sexes is large and broad and with the end-segment forms the **swimming fin.** The end-segment is called the **telson**; it bears no appendages; the **anal opening** is in its ventral side.

The natural color of the animal is usually a greenish black, but hot water or alcohol turns it red.

Exercise 1. Draw an outline of the dorsal side of the animal and label all the parts.

Cut off the right branchiostegite with the scissors, taking care not to injure the gills beneath. Push aside the gills and notice the thin integument which forms the lateral wall of the cephalothorax. Observe the method of attachment of the **gills.** They are feathery, thin-walled expansions of the body-wall and are attached either to it or to the basal portions of the legs. They present a very large surface to the surrounding water, and the blood circulating through them is thus oxygenated. Notice the **epipodites,** the skinny flaps which project from the basal joints of many of the legs and separate the gills of a segment from those of the next. They are not prominent in the crayfish.

Exercise 2. Without displacing the gills or epipodites make a sketch of them as they lie in the gill-chamber.

Exercise 3. Draw a diagram representing an ideal transverse section of the body-wall in the region of the walking legs; show the relations of the branchiostegites, the legs, and the gills to the body.

The appendages. Of these there are nineteen pairs, each somite of the body, with the exception of the last one, bearing a pair. There are thus thirteen cephalothoracic appendages, of which five are cephalic and eight thoracic, and six abdominal appendages. All of these appendages, except the first pair, however much they may differ from one another, are modifications of a single primitive type of structure. This type has been least modified in certain of the abdominal appendages. We shall, consequently, study these first.

Exercise 4. The abdominal appendages are called **swimmerets** or **pleopods**. Cut off the right swimmeret of the fourth abdominal somite close to the body, draw it on a large scale, and label all its parts. It consists of a basal piece, the **protopodite**, and two terminal branches, the inner or **endopodite**, and the outer or **exopodite**. This type of structure is characteristic of all crustacean appendages except the pair belonging to the first somite; those appendages which apparently differ from this type are modifications of it.

Exercise 5. Remove and draw on a large scale the right-hand sixth swimmeret. It is quite different from the last one drawn, and is sometimes called a **uropod**, but yet has the typical parts. Label its parts.

Exercise 6. (*a*) If the animal be a male, remove and draw the right-hand first and second swimmerets. These are modified from the typical structure to serve as copulatory organs.

Exercise 6. (*b*) If the animal be a female, remove and draw the right-hand first swimmeret.

Exercise 7. The five pairs of walking legs (including the chelipeds) are called **periopods** and belong to the thorax. Remove and draw the right-hand fourth periopod, disregarding the gill attached to it, and label the parts. It consists of seven segments, of which the two segments nearest the body constitute the protopodite, and the five farthest from the body the endopodite. The exopodite is not present.

Exercise 8. The **cheliped** is composed of the same segments as the other periopods. With a strong knife split the claw lengthwise into two equal halves. Examine the muscles controlling the movable limb of the claw. There is a strong **adductor muscle** which closes it, and a weaker **extensor muscle** which opens it. Make a diagrammatic drawing illustrating them.

Exercise 9. The three pairs of appendages directly in front of the chelipeds are the **maxillipeds**; they are thoracic appendages which assist in the process of eating. Remove with forceps and scissors the right-hand third (*i.e.*, the posterior) maxilliped; draw it on a large scale, disregarding the gill which may be attached to it, and carefully label the protopodite, exopodite, and endopodite.

Exercise 10. Remove with the forceps the right-hand second maxilliped and draw it on a large scale.

Exercise 11. Remove and draw the right-hand first maxilliped. The two large basal segments are the two segments of the leaf-like protopodite, the endopodite is a very small structure next to the protopodite, and the exopodite is a much longer structure next to the endopodite. A large epipodite is present. Label all of these.

Exercise 12. The two pairs of delicate appendages in front of the maxillipeds are the **maxillæ**; they are cephalic appendages. Remove with forceps the right-hand second maxilla and draw a large outline of it. The protopodite is wide and leaf-like; the endopodite is small; the exopodite is wide and with the epipodite forms a broad plate, called the **scaphognathite**, which is used by the animal to maintain a current of water from the gill-chamber.

Exercise 13. Remove carefully and draw the right-hand first maxilla. The protopodite is wide and leaf-like and similar to that of the second maxilla; the endopodite is extremely small; the exopodite is wanting.

Exercise 14. Extract with strong forceps and draw the right-hand **mandible.** The biting portion of it, together with the first joint of the small palp, forms the protopodite; the two terminal joints of the palp are the endopodite; the exopodite is wanting. Close to the posterior surface of the mandibles are the delicate **paragnatha.**

Exercise 15. Remove at its base and draw the right-hand second (the longer) **antenna.** The exopodite is a short, stiff, plate-like expansion; the endopodite is the long slender terminal portion.

Exercise 16. Remove and draw the right-hand **first antenna,** or **antennule** as it is also called. No exopodite and endopodite are present in this appendage throughout the Crustacea, there being typically but one terminal branch. In the crayfish and lobster this terminal branch is double.

Exercise 17. Construct in your notebook a table showing the relation of the appendages and somites, as follows:

No. of Somite.		Name of Appendage.	Prot.	Ex.	End.
Head ...	1				
	2				
	3				
	4				
	5				
Thorax ..	6	1st maxilliped	+	+	+
	7				
	8				
	9				
	10				
	11				
	12				
	13				
Abdomen .	14				
	15				
	16				
	17				
	18				
	19				
	20				

Write opposite the number of each somite, in your notebook, first, the name of the appendage belonging to it, then indicate what parts that appendage possesses by a " + " and what parts it lacks by a " — " under the appropriate head, as is shown above in the case of the sixth somite.

The gills. Remove the left-hand branchiostegite. Place the animal in water and study the gills on the left side. These organs may occur on the eight thoracic somites, and on each segment they may be attached either to the basal joint of the leg, when they are called **podobranchiæ**, to the flexible joint by which the leg articulates with the body, when they are called **arthrobranchiæ**, or to the body-wall just above the leg, when they are called **pleurobranchiæ**. A single thoracic somite may bear on each side four gills, — a pleurobranch, two arthrobranchs,

an anterior and a posterior, and a podobranch, — but on most of
the somites a less number is present.

Exercise 18. Construct in your notebook a table showing the
arrangement and number of the gills and also of the epipodites and their relations to the somites bearing them,
as follows:

No. of Somites.	Podo.	Ant. Arth.	Post. Arth.	Pleu.	Epip.	Total.
6						
7						
8						
9						
10						
11						
12						
13						

Begin with somite 13 and indicate by a " + " under the
proper head opposite the number of each somite the presence
of the gill or epipodite, and by a " —" its absence.

The internal organs. With strong scissors and forceps carefully remove the shell from the entire dorsal surface of the animal, taking great care not to disturb the organs lying beneath.
Notice just beneath the shell a pigmented membrane. This is
the **under-skin**; it is composed of a layer of connective tissue,
gland-cells, nerves, and blood, on the outer surface of which is
the layer of epithelial cells called the **hypodermis**, the matrix of
the shell. Entirely remove the under-skin. Study the organs
as they lie, without disturbing them. Notice in the cephalothorax, first, the large sac-like **stomach** just back of the rostrum
and connected by muscles with the anterior body-wall. On
each side of the stomach will be seen the cut ends of a mass of
muscle fibres. These are the **mandibular muscles**. Demonstrate
their connection with the mandibles. Just back of the stomach

is the white, shield-shaped **heart**, from the anterior end of which five delicate **arteries** proceed,—a median artery, and two pairs of lateral ones. Find these arteries and trace them forward as far as possible without breaking them. On both sides of the stomach and the heart and partly beneath them are the **liver** and the **reproductive organs**. The former is a pair of large, soft, and usually light-green organs which may fill a large portion of the cephalothorax and may extend back into the abdomen. The latter, if the animal be a female, are a pair of brownish or yellowish organs, the **ovaries**, in which the ova can often be seen; they are situated beneath the heart and in front of and behind it, and vary in size and also in color with the development of the ova. When these are approaching maturity the ovaries are the most prominent organs in the body-cavity, and often extend far back into the abdomen. In the male animal the reproductive glands, the **testes**, are white in color and very slender, and occupy the same position as the ovaries in the female. Note the coiled **vas deferens** on each side.

Study the musculature and the other organs of the abdomen. There are two systems of muscles here. On the dorsal side are longitudinal muscles, the **extensors**, which extend or straighten the abdomen. Separate these muscles carefully along the median line and observe beneath them the delicate, colorless **abdominal artery** which carries the blood from the heart throughout the abdomen. Trace it forward to the heart. Notice the lateral **branch-arteries**. How many pairs are there? Just beneath this artery lies the **intestine**, which often contains dark-colored fæcal matter. Beneath it and filling most of the space within the abdomen are the **flexor muscles**, which are very complex, whose function it is to bend or flex the abdomen. It is by the use of these two sets of muscles that the animal swims.

Exercise 19. Draw an outline of the animal's body, showing the segmentation and the above-mentioned organs *in situ*, and label them all.

The circulatory system. The **heart** lies within an enclosed space called the **pericardial sac**, the walls of which, the **pericardium**, will have been partially destroyed by the removal of the under-skin. The heart, the abdominal artery with its lateral branches, and the five anterior arteries have been studied and drawn. Carefully press aside the heart and note the median **dorso-ventral artery** which leaves the abdominal artery near the heart and passes ventrally. This artery supplies with blood a **ventral longitudinal artery**, which lies in the mid-ventral line in the thorax and abdomen.

Remove the dorsal abdominal artery and the heart from the body and float them in clean water. Note the six valvular openings of the heart, two being on the dorsal side, two on the ventral, and one on each of the lateral sides. These can be seen by blowing on the heart through a blow-pipe.

Exercise 20. Draw a dorsal view of the heart showing the valves there present.

The course of the circulation of the blood is the following: by the contraction of the heart the blood is sent through the arteries to all parts of the body; after bathing the different tissues it collects in a ventral blood-sinus, a passage in the ventral portion of the body-cavity in which lie the ventral nerve-chain and the ventral abdominal artery, and passes towards the gills; from the ventral sinus it passes to the gills through afferent veins, one of which runs to each gill and along the outer edge of it; it then runs through the delicate gill-filaments, where it is aërated, and passes by efferent veins on the inner edges of the gills back to their base; here six larger branchial veins collect the blood and carry it to the pericardial sac, whence it is taken through the valvular openings into the heart.

Exercise 21. Draw a diagram representing the entire circulatory system.

The reproductive system. The female genital organs. The position of the **ovaries** has already been observed. In the crayfish their forward portions are paired, while their hinder portions are fused and lie in the median line. In the lobster, however, no such fusion takes place, but the two ovaries are united by a bridge midway in their length. Find the paired **oviducts** which lead from the ovaries to the genital openings. Remove both ovaries and oviducts from the body and float them in water.

Exercise 22. (*a*) Make a diagrammatic sketch of them.

The male genital organs. The position of the **testes** has been already noted. In the crayfish they are similar in shape and position to the ovaries in the female animal, but are more slender ; in the lobster they are a pair of long white tubes which extend forward as far as the stomach and back into the abdomen. Find the paired **vasa deferentia,** which are long convoluted tubes connecting the testes with the external genital openings. Remove the vasa deferentia with the testes from the body and float them in a pan of water.

Exercise 22. (*b*) Make a diagrammatic sketch of the male reproductive tract.

Cut open a vas deferens and examine its contents under a high power of the microscope. Star-shaped **spermatozoa** will be seen.

Exercise 22. (*c*) Draw a spermatozoan.

The **digestive tract** consists of the **mouth, œsophagus, stomach, intestine,** into which open the paired **livers,** and the **rectum.** Pass a probe through the mouth into the stomach and notice the dorso-ventral course of the œsophagus, which joins the mouth with the stomach. The paired ducts which unite the two lobes of the liver with the intestine join that organ just back of the stomach. Find them. With scissors sever the œsophagus just ventral to the stomach, taking care not to injure the brain, which lies in front of the stomach, or the two slender nerve-connectives, which lie on either side of the œsophagus. Sever the rectum near the anus. Remove the entire digestive tract from the body and place it in a pan of clean water. The liver is so soft that it may not be possible to remove it entire. Notice the boundary between the intestine and the somewhat larger rectum. In the crayfish the rectum is much longer than the intestine; in the lobster the opposite is true. In the lobster notice the **blind-gut** or **appendix** which joins the rectum near its anterior end.

Exercise 23. Make a diagrammatic sketch of the digestive tract.

Cut open the stomach by a ventral incision and wash it out. Observe its chitinous lining and the dark brown chitinous teeth. This chitinous lining is a continuation of the cuticula which covers the external surface of the body and is moulted with the cuticula. During certain parts of the year a pair of large calcareous bodies called **gastroliths** are imbedded in the lining of the stomach. They remain in the stomach after the moulting of the cuticula and furnish lime for the new cuticula, which at once grows rapidly.

Exercise 24. Make a sketch of the inner surface of the stomach showing the teeth.

The excretory system. Notice in the extreme forward end of the body-cavity, just in front of and below the stomach, a pair of pale greenish bodies. These are the kidneys or **green glands**. Each one is made up of two portions, the smaller glandular portion, next to the body-wall, and the larger saccular portion, or urinary bladder, next the stomach. From the latter the **ureter** leads to the external openings which have already been noted.

Exercise 25. Draw a view of the forward end of the body-cavity showing the kidneys as they lie in position.

The nervous system consists of a ventral double **nerve cord** lying in the mid-ventral line in the body-cavity and extending the length of the animal, with paired **ganglia** at intervals, also of a **brain** situated just back of the eyes, which is united with the ventral nerve by two **nerve connectives**, passing one on each side of the œsophagus. The ventral ganglia have typically a metameric significance, but many of the somites have lost their ganglia, so that there are fewer ganglia than somites. The double nature of the ventral nerve is best seen in the thorax.

Remove all the muscles and the viscera from the body. The ventral nerve cord will be seen in the abdomen lying in the mid-ventral line. Notice the ganglia. How many do you count? Notice the lateral nerve-branches. In the cephalo-thorax the nerve cord is concealed beneath transverse ridges of the ventral wall of the shell. Cut these with scissors and expose the nerve, beginning at the hinder end of the cephalo-thorax and working forward. How many thoracic ganglia do you find? Just back of the œsophagus is the large **subœso-phageal ganglion** which is connected with the brain by the two connectives already mentioned. The **brain** or **supraœsophageal ganglion** is just back of the eyes.

Exercise 26. Draw an outline of the body and in it the nervous system, showing accurately the number of ganglia and the segments in which they lie, together with the lateral nerves.

Exercise 27. Remove the brain and draw an outline of it on a scale of 6 or 8, using a dissecting microscope or hand lens. Show the antennal and the optic nerves.

Exercise 28. Draw a diagram representing an ideal sagittal section of the animal in which the relative position of the principal systems of organs is accurately shown.

CRUSTACEA

A BRACHYURAN DECAPOD. A CRAB

The crab is a representative of the more highly specialized of the two divisions of the Decapoda, the Brachyura, which include those decapods with short weak abdomens. The lobster and the crayfish represent the other and less highly specialized of the two divisions, the Macrura, which comprise those decapods with long abdomens.

Compare the crab with the lobster or the crayfish. Note the broad shield-shaped **cephalothorax** and the **abdomen** bent under it. The abdomen of the male crab is narrow while that of the female is broad. Which sex is your animal? In what ways is the higher specialization of the cephalothorax and the abdomen of the crab shown?

The body of the crab is composed of twenty somites, like that of the crayfish and the lobster, thirteen of which belong to the cephalothorax and seven to the abdomen. The cephalothorax is covered by a **carapace**. Notice the short transverse suture which separates the cephalic from the thoracic portion. At the ends of this suture notice the longitudinal depressions which mark off the lateral branchial areas and separate the **branchiostegites** from the median portion of the carapace. The branchiostegites are not applied closely to the body as they are in the lobster and the crayfish, but stand out from it, very much increasing the transverse axis of the cephalothorax and making it longer than the longitudinal axis. This feature of its structure makes it easy for the crab to run sideways. Notice that the ventral edge of the branchiostegite is closely applied to the body, so that the respiratory water could hardly enter the gill-chamber along this edge as it does in the crayfish and the lobster. An opening is present, however, at the base of the cheliped through which the water enters. Pass a probe into the branchial chamber through this opening. Notice the prominent stalked **eyes**; also the two pairs of delicate **antennæ**. Examine and identify the **mouth-parts** and the **thoracic legs**; they will be found to correspond to those of the crayfish or the lobster. Find the openings of the genital organs; in the male on the ventral surface of the last and in the female of the antepenultimate cephalothoracic segment.

The abdomen is relatively small and weak and usually remains folded beneath the cephalothorax. It lacks the swimming fin; most crabs cannot swim. The common blue crab, however, swims very well by means of the fifth pair of periopods. The number of abdominal segments is variable, fusion having taken place between certain of the somites. This number is also not the same in the male and the female of the same species. Raise the abdomen from the cephalothorax and observe the **swimmerets** on its ventral surface. In the female note the long chitinous hairs which fringe the swimmerets. It is to them that the eggs and newly born young are attached. The only swimmerets present in the male are the first two pairs, which are functional in pairing.

Exercise 1. Draw a dorsal view of the animal with the abdomen extended, being careful not to omit the antennæ and the eyes, and label all the parts observed.

Exercise 2. Construct in your notebook a table showing the relation of the appendages and somites similar to that made use of with the lobster or the crayfish. (See page 35.)

The gills. With stout scissors cut off the right branchiostegite and expose the gills. These will be found to be quite different from those in the lobster or the crayfish, **pleurobranchiæ** only being present. Note the enormously elongated **epipodite**

of the first maxilliped which extends across the gills to the hinder part of the branchial chamber.

Exercise 3. Construct a table showing the relation of the gills to the somites similar to that made use of in the dissection of the lobster or the crayfish.   (See page 36.)

Exercise 4. Draw a diagrammatic cross section representing an outline of the body-wall in the region of the walking legs; show in this the relation which branchiostegites, legs, and gills bear to the body.

Internal anatomy.   With strong scissors and forceps remove the shell from the entire dorsal surface of the body, taking care not to injure the organs within.   The arrangement of the organs will be seen to be similar to that in the crayfish or the lobster. The **livers** are a pair of extensive yellowish organs.   The anterior portion of each of these passes laterally into the cavity of the branchiostegite; the posterior portion passes backward beneath the **heart**.   In the male animal the **testes** are whitish organs which follow the course of the livers; the **vasa deferentia** are slender, coiled tubes which lie on each side of the heart. In the female animal the **ovaries** also accompany the livers; the **oviducts** are a pair of tubes which pass to the genital openings, the middle portion of each being expanded to form a large sac, the **receptaculum seminis.**

Exercise 5. Draw an outline of the body and the organs as they lie *in situ*.   Label all carefully.

Remove all the viscera, taking care not to injure the brain and the circumœsophageal nerves, and examine the **nervous system.**   The **brain** is just back of the eyes, as in the lobster or the crayfish, and is united with the ventral nerves by means of the lateral **circumœsophageal connectives** which pass on each side of the œsophagus.   There is, however, no long ventral nerve cord with segmental ganglia, but a single **large ganglionic** mass, in the shape of a ring, which occupies a central position in the cephalothorax, and from which nerves radiate to the different appendages. The dorso-ventral artery passes through this ring. Expose the entire nervous system.

Exercise 6. Draw a semidiagrammatic view of the nervous system, being careful to represent accurately the nerves radiating from the ganglionic ring and those going from the brain to the eyes and to the antennæ.

CRUSTACEA

A LAND ISOPOD. A SOW-BUG (*Porcellio, Oniscus,* or *Armadillidium*)

This animal is one of the few terrestrial crustaceans. It may be found at any time of the year under stones, logs, etc., and in other moist, dark places, where it lives on decaying vegetable matter.

The animal must be studied with the aid of a hand lens or a dissecting microscope. Compare the animal with the crustaceans already studied. Notice the flattened body. It is composed of twenty somites, of which five are cephalic, eight are thoracic, and seven are abdominal, and much less fusion has taken place among them than is the case in the decapods. The head and the thorax are not covered by a carapace and thus are not joined together to form a cephalothorax. The apparent head is composed of six fused somites, five of which are cephalic and one thoracic. The remaining seven thoracic somites are free and movable. Count them. Count the abdominal segments. Six will be found, the last two abdominal somites being fused together.

Find the **eyes**: they are not on stalks, but are sessile. Only one pair of **antennæ** appears, the first pair being rudimentary. Notice the pair of **anal feelers** which extend back from the hinder end of the body. These are homologous to the last pair of appendages, like the cerci of orthopterous insects, and have a similar function.

Exercise 1. Draw a dorsal view of the animal on a scale of 10. Number the thoracic and the abdominal segments.

Study the ventral side of the animal. Notice if it be a male or a female. The male has a long dark-colored, tube-shaped

copulatory organ which extends from the forward border of the abdomen backward. The female, besides lacking this organ, may have a **brood-sac** on the ventral surface of the thorax, which is composed of plates attached to the inner side of the first five pairs of walking legs and contains eggs or young.

The appendages. First observe the seven pairs of **walking legs**; they are the thoracic legs numbering from two to eight; exopodites and gills are wanting in them. The **gills**, instead of being thoracic structures, as in the decapods, are attached to the abdominal legs. With a fine needle separate the flattened appendages of the first five abdominal segments. The endopodite serves as the gill, while the exopodite is large and plate-like and covers the endopodite. The appendages of the head may be best studied from the hinder pair forward. They consist of one pair of **maxillipeds**, which belong to the first thoracic somite, two pairs of **maxillæ**, one pair of **mandibles**, and one pair of **antennæ**, the second, the first pair of antennæ being rudimentary. The maxillipeds are plate-like and cover the other mouth-parts. Carefully remove the maxillipeds and study the mouth-parts.

Exercise 2. Construct a table showing the relation of the appendages and the somites similar to that made use of in the dissection of the crayfish or the lobster (page 35), leaving out of consideration, however, the protopodites, exopodites, and endopodites.

CRUSTACEA

A TYPICAL AMPHIPOD. A FRESHWATER SHRIMP (*Gammarus*)
OR A SAND-FLEA (*Talorchestia*)

The freshwater shrimp is common in many places in pools and streams, and may be easily caught with a fine net; the sand-flea is a marine animal and is extremely common along all of our shores.

Notice the compressed and translucent body; this latter feature is extremely wide-spread among the smaller aquatic animals. Can you explain what is the advantage to a small aquatic animal to be translucent or transparent? Note the two pairs of long antennæ. In common with all the higher crustacea, the body is composed of twenty somites, of which five are cephalic, eight thoracic, and seven abdominal. Like the isopod, the animal has no carapace, the eyes are sessile, and the apparent head is composed of six fused somites, five being cephalic and one thoracic. There are thus seven free thoracic segments. Note the broad movable plates, the **epimeral plates**, which depend from the ventral side of certain of the thoracic segments, extending the lateral surface of the body ventrally; note the differences in form between the thoracic appendages. The abdomen is composed of six free segments, the sixth and seventh somites being fused. Count them. The first three pairs of abdominal legs are swimming legs, the last three are jumping legs.

Exercise 1. Draw an outline of the side view of the animal on a large scale. Number the thoracic and the abdominal segments.

Study the appendages, beginning with the free thoracic ones. With fine needles separate the legs and observe the **gills** attached to the posterior borders. How many bear gills? In the female observe the **brood-sac** when it is present ; it is formed by plate-like projections of the inner side of certain thoracic feet. In the abdomen observe the **biramous appendages** ; they bear no gills. The cephalic appendages are those typical of crustacea. In front of the mouth is a median lip called the **labrum**, which, however, is not an appendage. Then come the **mandibles** and two pairs of **maxillæ**. The pair of appendages, the **maxillipeds**, belonging to the first thoracic somite (which is fused with the cephalic somites) form a kind of **lower lip**.

Exercise 2. Construct a table of somites and appendages similar to that made use of in the dissection of the lobster or the crayfish (page 35), leaving out of consideration, however, the protopodites, exopodites, and endopodites.

CRUSTACEA

AN ABERRANT AMPHIPOD (*Caprella*)

This is a very common marine amphipod which is found along our shores clinging to hydroid colonies and to seaweed. It is an interesting form because it illustrates an extreme degree of modification from the typical amphipod type; a modification which is the result of its peculiar environment.

Notice the irregular cylindrical form and the small number of appendages. The apparent head is composed of seven fused somites, of which five are cephalic and two are thoracic, the first of these latter bearing a pair of **maxillipeds**, and the second a pair of **legs**. There are thus six free thoracic segments, of which the first, fourth, fifth, and sixth bear **non-branchiate legs**, and the second and third bear **gills** but no legs. The abdomen has lost its segmentation and its appendages and has been reduced to a mere protuberance at the end of the thorax.

Exercise 1. Draw a large outline of the side view of the animal. Number the segments and label the parts observed.

Exercise 2. Construct a table of somites and appendages similar to that made use of in the dissection of the crayfish or the lobster (page 35), leaving out of consideration, however, the protopodites, exopodites, and endopodites.

CRUSTACEA

LARVAL DECAPODS: THE ZOËA OF THE CRAB; THE MEGALOPA OF THE CRAB; THE MYSIS STAGE OF THE LOBSTER

These names have been given to certain larval forms of the crab and the lobster, as well as to those of other of the higher crustaceans. It is as **zoëæ** that the crab and the higher crustaceans generally leave the egg. The zoëa of the crab grows into the **megalopa,** which in time grows into the adult animal. The stage in which the lobster is born is more advanced than the zoëa and is called the **mysis stage.** All of these larvæ are minute animals and are more or less common in the surface waters of the sea along our coast.

Mount several **zoëæ** of the crab on a slide and study them under the microscope. The body will be seen to be divided into two body-divisions, a **cephalothorax** and an **abdomen.** The former is covered with a delicate **carapace,** from which project one or more spines. When the animal is newly born it possesses the typical five pairs of cephalic appendages, and the anterior two or three pairs of thoracic appendages, *i.e.,* the maxillipeds, which, however, are used for locomotion. The remaining thoracic and the abdominal appendages are wanting, but appear as the animal increases in size, those anteriorly situated appearing first. The animal has two stalked eyes.

Exercise 1. Draw a side view of a zoëa on a large scale, representing accurately the appendages, and label the parts observed.

Mount a **megalopa** and study it under the microscope. We observe that it is much larger than the zoëa, that it has acquired a relatively much larger cephalothorax and abdominal appendages, and is much more crab-like than the zoëa. But it still has a long abdomen, and at the end of this is a swimming fin. The megalopa is a swimming animal, like the adult lobster, but it is gradually assuming the characters of the adult crab. Its two anterior maxillipeds have lost their locomotory character, which they possessed in the zoëa, and have assumed their final form and function. Identify all the mouth-parts.

Exercise 2. Draw a dorsal view of the animal, with the legs extended, on a large scale.

Mount several **lobster larvæ** in the mysis stage and study them under the microscope. The lobster is born in a more advanced condition than is the crab. The zoëa stage of the lobster is passed over in the egg, and when the young animal emerges from the egg it resembles Mysis, a schizopodous crustacean, and hence is said to be in the mysis stage. The general form of the animal does not differ much from that of the adult. The abdomen bears no appendages. The cephalothorax is very nearly like that of the adult and bears the same appendages. The third maxilliped, however, is a locomotory appendage, as it is in Mysis, and with the five periopods is used for swimming. Notice the biramous character of each periopod.

Exercise 3. Draw a side view of the animal on a large scale.

CRUSTACEA

A FREE-SWIMMING COPEPOD (*Cyclops*)

These minute animals are representatives of the division of
Crustacea called the Entomostraca. All of the crustaceans
heretofore studied belong to the higher group called Malacostraca. Copepods are extremely common in both fresh and salt
water. They may be obtained in almost any permanent pool
of water in the woods or fields or from the surface water of the
sea, often in large quantities, and are easily kept in aquaria.
The animals should be studied alive if possible. Place several
on a slide under a cover-glass and examine them under a microscope. If the pressure of the cover-glass does not suffice to
keep them quiet, the withdrawal of some of the water from
under the cover-glass with blotting-paper will probably accom-
plish this result. Also stain and mount a number of copepods
in balsam or glycerine. Observe the cylindrical body and the
two pairs of long **antennæ** with their sense-hairs; also the long
spines at the end of the abdomen. Note the division of the
body into **abdomen** and **cephalothorax,** and also that the latter is
not covered by a carapace. If the animal be a female it may be
carrying a pair of **egg-sacs** filled with eggs extending from the
anterior end of the abdomen. Note the **median eye,** also the
intestine and **muscle fibers,** through the transparent body-wall.

The body is made up of fifteen somites, the **head, thorax,** and
abdomen each containing five. The head is relatively large, and
its somites are fused together; they bear the cephalic appendages common to all crustaceans. The first pair of **antennæ** is
longer than the second; in the male it is secondarily modified

to form clasping organs, by which the female is held during pairing. In **Cyclops**, which is the commonest freshwater genus of the Copepoda, the first thoracic somite is fused with the head, leaving only four free thoracic somites. The abdomen bears no appendages. In the female the first two abdominal somites may be fused together.

Exercise 1. Draw a large outline of the dorsal aspect of a copepod, not putting in any appendages except the antennæ. Represent accurately the sense-hairs on the antennæ and the caudal bristles. Number the thoracic and abdominal segments and carefully label all the parts.

Study the appendages. The thoracic appendages are biramous. They do not bear gills, and the fifth pair is rudimentary. The cephalic appendages consist of two pairs of **antennæ**, one pair of **mandibles**, and two pairs of **maxillæ**, the second pair of which are without protopodites. The exopodites and endopodites of this second pair join the body separately in consequence and may appear as independent appendages.

Exercise 2. Draw a side view of an animal showing the appendages in position.

Exercise 3. Draw an outline of a thoracic leg on a large scale, showing accurately all the joints and hairs.

Compare the copepod with the young larva of the crab or the lobster. Enumerate the points of structural similarity between them.

Internal anatomy. This can be best studied in the live animal. The **alimentary tract** is straight and of large diameter, and often contains dark-colored fæcal matter. The **mouth** has a ventral position, as in other crustaceans, while the **anus** is dorsal. There is no liver or other accessory glandular organ. The **circulatory system** in Cyclops consists of the colorless blood fluid alone, there being no heart. The blood is, however, kept in circulation by the rhythmic contractions of the intestine. Other copepods possess a dorsal heart. There are no special respiratory organs. How is respiration carried on ? The **excretory system** consists of a pair of coiled tubes, called the **shell glands,** which lie in the forward part of the cephalothorax and have external openings near the base of the first pair of thoracic appendages.

The **reproductive system** consists of median or paired organs in the dorsal portion of the cephalothorax above the intestine. In the female the **ovaries** are often conspicuous as a pair of large branched organs. The **oviducts** are paired and lead to the exter-nal sexual openings in the first abdominal segment. Appended to the first abdominal segment may be a pair of **egg-sacs** containing fertilized eggs which are cemented together by means of a secretion of the oviduct. In the male the reproductive gland is the median **testis,** which communicates by means of paired **vasa deferentia** with the external sexual openings, which are also in the first abdominal segment. The spermatozoa collect in the terminal portion of each vas deferens and form there a small mass known as a **spermatophore.** The two spermatophores, during the act of pairing, pass to the female and fertilize the ova. The male animals are much less numerous than the females.

The reproductive glands of the copepod can be observed as above described only during times of sexual activity. At other times they can be seen only in part or not at all.

The **muscular system** can be easily seen to consist of striated muscle fibers. Longitudinal as well as converging fibers will be seen at each appendage.

The **nervous system** may be seen in favorable specimens as a ventral strand in the cephalothorax connecting with the large dorsal **brain.**

Exercise 4. Draw a side view of the animal, showing as many of the internal organs as you have observed.

CRUSTACEA

A CLADOCERAN PHYLLOPOD (*Daphnia*)

This is a small freshwater crustacean common in lakes and pools. It should be studied under the microscope and alive if possible. Place several on a slide under a cover-glass and draw off enough water to keep them quiet; also observe several in a watch-glass in order to see them from above. The body of the animal will be seen to differ in shape from those crustaceans already studied. It is but indistinctly segmented, and, except the head, is entirely covered by a bivalve shell. This shell is the cuticular covering of paired folds of the dorsal integument, one fold covering each side of the body. Beneath the opening of the valves of the shell appear the **appendages** and the **abdomen**; on the surface of the shell a mesh-work of fine lines can usually be seen. Notice the large, median **eye**; it may often be seen to tremble slightly. The shell has a deep, ventral indentation near the base of the antennæ.

The **first pair of antennæ** is very small, but may be easily seen projecting downward just back of the eye. The **second pair of antennæ** is very long and biramous, the two branches being the exopodite and endopodite; they are the principal organs of locomotion. Just back of the antennæ is a large flap, called the **upper lip**, and back of this are the large **mandibles**. There is but a single pair of **maxillæ**, and they are so small that they will probably not be seen. Four to six pairs of **thoracic appendages** follow, the function of which is probably exclusively respiratory. How many are present in your specimen? Notice the leaf-like surface of these appendages (whence the name phyllopod), in which we can recognize the basal protopodite and two broad terminal pieces, the endopodite and exopodite. The short abdomen articulates with the thorax and is bent beneath it, where it may be seen often moving rapidly back and forth.

Exercise 1. Draw an outline of a side view of the animal on a large scale and label the appendages and other parts observed.

Exercise 2. Draw a highly magnified view of one of the thoracic appendages and label accurately all the parts.

Internal organs. The **digestive tract** passes from the mouth, which is ventrally placed and lies back of the ventral cleft in the shell, first forward, then turns dorsally and finally posteriorly and extends back to the **anus** near the end of the abdomen. Near the anterior bend of the digestive tract a pair of colored, curved pouches communicate with it ; they are **liver-sacs.** The sac-like **heart** may be seen beating rapidly above the intestine. It possesses a pair of lateral openings into which the blood streams from the body-cavity with each dilation, and an anterior opening through which it is sent into the forward part of the body with each contraction. There are no other blood vessels. Below the heart is a pair of excretory glands, called the **shell glands,** which open to the exterior near the mouth. The **nervous system** consists of an **optic ganglion** and a **brain,** lying back of the eye, and a **ventral nerve** containing seven **ganglia.**

The reproductive organs. The daphnias which are usually seen are all parthenogenetic females, the males making their appearance at certain times of the year only. The female animal is larger than the male, and may be distinguished by its **brood-sac.** This is a large space just beneath the dorsal wall of the thorax in which the eggs and the young brood are carried. The **ovaries** are a pair of tubular organs alongside the intestine, which communicate, by means of short **oviducts,** with the brood-sac. The ovaries are easily detected by the presence of large ova in them. These are in groups of four, of which but one, the third, is destined to become an egg, the other three being nutritive cells by which it is nourished. In the male the **testes** occupy a position similar to that of the ovaries. Their external openings are on the ventral side of the abdomen.

During the greater part of the summer the eggs pass into the brood-pouch unfertilized and develop there parthenogenetically, producing only females. The young animals pass out of the brood-chamber through a posterior opening ; they soon become adult and in their turn give birth to parthenogenetic females. The eggs which thus develop are called **summer eggs**. At certain times of the year, however, as in the autumn, males are also born. They fertilize the females, and the fertilized eggs then produced differ from those which were unfertilized in possessing thicker shells. They are called **winter eggs** and are able to resist the cold of winter or the effect of drought. In the springtime the winter eggs develop into parthenogenetic females again.

Exercise 3. Draw an outline of the animal and place in it all the internal organs you have observed.

CRUSTACEA

A LARVAL ENTOMOSTRACAN. A NAUPLIUS LARVA

In an aquarium containing copepods or ostracods there are sure to be numbers of the young larvæ of these animals. They are minute, free-swimming forms and are called **nauplii**, and may be recognized by the triangular or oval, unsegmented body, which bears three pairs of appendages and a median eye. Nauplii of marine entomostracans may also be met with in large numbers among the small animals obtained by skimming the surface waters of the sea with a fine net.

Examine in a watch-glass under a microscope water containing sediment taken from a jar in which are copepods or ostracods. Find a nauplius; the ostracod nauplius differs from that of the copepod by being enclosed in the characteristic bivalve ostracod shell. If marine plankton is at hand, look for several kinds of nauplii in it.

Study the structure of a nauplius. Observe the unsegmented body; if the animal is not newly born, signs of segmentation may have begun to appear. Observe the three pairs of segmented appendages ; the segmentation, however, is often indistinct. These appendages are homologous to the **first** and **second pair** of **antennæ** and the pair of **mandibles** of the adult animal. As in the adult, the first pair is uniramous; the second and third pairs are biramous. Both of the latter two pairs are used for locomotion, although it is probable that they also act as jaws. The **median eye** will be seen, and the straight **digestive canal.**

Exercise 1. Draw a nauplius on a large scale and label all the parts above mentioned.

The nauplius larva is of great theoretical significance. It appears as the youngest, free-swimming larval stage of almost all the entomostracans and of several of the malacostracans, and those malacostracans which are born in a later period of their development pass through a nauplius stage (*i.e.*, a stage in which the body is unsegmented and bears three pairs of appendages) while they are still in the egg. This universal occurrence of the nauplius larva seems to indicate that it repeats substantially the structure of the primitive ancestor of all crustaceans.

In its further development and growth the nauplius larva increases in size, gradually becomes segmented, and acquires new appendages, its growth and the specialization of its organs advancing from the anterior towards the posterior end. The appendages, which were originally typical, unmodified crustacean appendages, become differentiated to form the first and second pairs of antennæ and the mandibles, and finally the size and structure of the adult are attained.

Exercise 2. Look for several nauplii which are somewhat advanced in development and draw outlines of them.

CHAPTER II

ANNELIDA

A POLYCHAETOUS ANNELID (*Nereis virens*)

Nereis is a common marine worm which. lives in the sand along the shores of our northern and middle states. Its food consists of various kinds of small marine animals, which it catches with its formidable, protrusile proboscis. A specimen should be selected for study in which the proboscis is not thrust out.

Observe, in the first place, the long, segmented, and somewhat flattened body, the pair of appendages on each segment, and the distinct head with special sense-organs at the forward end; observe also that the body tapers towards the hinder end, where is a pair of special sense-organs, the long caudal feelers. All of these characters indicate an animal possessed of the power of rapid locomotion. Count the **somites** or body-segments; note that they are almost exactly alike. This lack of specialization is in sharp contrast to the condition of the somites in most arthropods. Observe carefully the appendages; they differ from those of the arthropod in that each one is an unjointed expansion of the body-wall, whereas the arthropodous appendage is segmented. Each one is made up of several lobes and is provided with long bristles or **setæ**. Note the absence of a hard shell, the external integumentary covering being the glistening cuticula, which has not been stiffened by the presence of calcareous salts.

Observe the head and the forward portion of the body. An annelid's body is composed genetically of two portions:

the **prosoma**, or primitive head, and the **metasoma**, or the primitive segmented trunk. The prosoma may be further divided into the **prostomium**, which lies in front of the mouth and contains the brain and the principal organs of special sense, and the **metastomium**, which contains the mouth. In Nereis the prostomium bears the following special sense-organs: a pair of **palps**, large cylindrical projections extending forward at its anterior end; a pair of **tentacles**, two delicate organs between the palps; and two pairs of **eyes**, small bead-like organs near the base of the palps. Carefully identify all of these organs and notice whether the palps and tentacles are jointed. The metastomium is, in Nereis, fused with the first two somites of the metasoma or trunk, and the segment thus formed is called the **peristomium**. It bears the mouth and four pairs of long, flexible sense-organs called the **peristomial cirri**. Carefully observe, with the aid of a hand lens, their exact position. These cirri are morphologically not cephalic organs, as are the palps and the tentacles, but are remnants of appendages of the first two somites.

Exercise 1. Make an outline of the dorsal aspect of the head and the first five or six somites on a scale of 5. Number the somites. Carefully label all the parts.

Exercise 2. Draw a side view of the head on a scale of 5. Take special care to represent accurately the position of the peristomial cirri.

Exercise 3. Find a specimen, if possible, with the proboscis thrust out and draw a dorsal view of its head.

Note the tapering of the body at the hinder end. The worm grows in length at this end. The posterior somites are the youngest and hence the smallest.

Exercise 4. Make a sketch of the hinder end of the animal. The long sense-organs at the extreme end are called **caudal cirri**. In which direction do they project?

The appendages in annelids are called **parapodia**. Carefully examine the parapodia at different parts of the body and see if they are all alike.

Remove a parapodium from the middle of the body; mount it on a slide in glycerine or water and study it with the aid of a hand lens or a microscope. Compare it with the parapodia still on the animal and determine which is its dorsal and which its ventral side. It can be divided into two distinct portions, the dorsal and the ventral portions, called the **notopodium** and the **neuropodium**, respectively, each of which is stiffened by an internal chitinous supporting rod, called the **aciculum**. Find the two acicula. The large dorsal lobe of the notopodium is a respiratory organ, **a gill**. It contains branching blood vessels which can be easily seen. Attached to its dorsal edge is a slender, vibratile sense-organ, the **dorsal cirrus**. Beneath the gill are two lobes, one bearing bristles or **setæ**. The neuropodium is made up of two lobes, one of them setæ-bearing, beneath which is a **ventral cirrus**.

Exercise 5. Draw a parapodium on a scale of 6 and label the parts.

Remove a parapodium from the hinder end of the animal, mount it, and study it. Has it the same parts, and if not, which are missing?

Exercise 6. Draw it on a scale of 6.

Internal anatomy. Make an incision with fine, sharp scissors in the mid-dorsal line of the integument of the anterior third of the animal, taking great care not to injure the viscera which lie beneath. The body will be seen to be divided into compartments corresponding to the somites, by transverse partitions which are called **septa**. Holding the cut edge of the integument with forceps, cut the septa where they join it, and then spread out and pin down the body-wall, using many pins on each side.

The digestive organs. The **mouth** leads into the large **pharynx,** which is composed of an anterior and a posterior portion. With sharp scissors cut open the pharynx along the mid-dorsal line and note the number and arrangement of the chitinous **teeth** imbedded in its inner surface. Notice the delicate muscles passing from it to the body-wall by means of which the pharynx can be thrust out of the mouth and drawn back again. They are the **protractors** and the **retractors.** A pharynx which is thus protrusile is called a **proboscis.** Just back of it is the narrow **œsophagus** with which a pair of small tubular **glands** communicates. Back of the œsophagus is the **stomach-intestine,** which extends to the **anus.** Observe the **mesenteries.** These are longitudinal partitions, in structure like the septa, one of which attaches the stomach-intestine to the dorsal and the other to the ventral body-wall. Press the intestine aside and see the ventral mesentery.

The circulatory system. Nereis has two distinct circulatory fluids, the colorless or cœlomic and the red blood fluid. The first consists of a plasma in which float amœboid blood cells; it circulates freely in the body-cavity or cœlom, being forced by the movements of the body from one segment to another through small openings in the septa. The red blood consists of a red plasma, in which float colorless blood cells, and circulates in closed tubes. The most important of these blood vessels are two longitudinal tubes, the dorsal and the ventral arteries, which lie in the median line, one above and the other below the alimentary canal. The former, **the dorsal artery,** pulsates and drives the blood towards the forward end of the body and distributes it to lateral **segmental arteries.** Observe these and determine how many there are in each segment; also note the capillary **network** into which the dorsal artery breaks up at its anterior end. The dorsal portions of the lateral arteries carry the blood to the gills and other organs, whence it collects again in the ventral portions of these arteries and is conducted to the ventral artery. In this vessel the blood flows toward the hinder end of the body.

Exercise 7. Draw a view of the opened animal on a scale of 5, showing the organs above described. Label all the organs carefully.

Sever the alimentary tract at the œsophagus and remove the stomach-intestine from the body. Observe the **muscle bands** in the body-wall; note the difference in direction and size of the different bands. Observe the muscles at the base of the acicula.

The excretory system. The **kidneys** of the animal consist of a pair of glandular organs called **nephridia**, which lie in the body-cavity against the ventral body-wall in each somite except the last two or three. Each nephridium opens through the body-wall to the exterior in a minute pore on the ventral surface of each somite near the base of the parapodium. The anterior end of the nephridium passes through the septum which forms the anterior wall of the somite in which that organ lies, and opens into the body-cavity. The opening, which is ciliated, is called the **nephrostome**; it lies, as will be seen, against the anterior surface of a septum. Study the nephridia carefully in several parts of the body under a dissecting microscope; some of them may have been torn in removing the intestine. Examine a portion of the worm in which that organ is still in the body and note the relation of the nephridia to it and to the septa.

Exercise 8. Draw a diagram representing the opened body-cavity in a number of somites and the position of the nephridia and the muscles.

The nervous system. Observe in the mid-ventral line of the body-cavity the **nerve cord.** Trace it forward to the **brain.** Note the **connectives** which encircle the pharynx and connect it with the brain. Remove the forward end of the nervous system from the body, mount it in glycerine, and study it under a microscope. Note the **ganglionic enlargements** and the double nature of the nerve cord. Study the branches which pass off from the nerve cord and from the brain.

Exercise 9. Draw the nervous system on a scale of 10, showing all the features above mentioned.

The reproductive system. There is no complicated reproductive system in Nereis. The sexes are separate. The reproductive glands make their appearance only during the periods of sexual activity and then as swellings of the peritoneal lining of the body-cavity. The eggs or spermatozoa, as the case may be, fall into the body-cavity and find their way to the outside through the nephridia or through temporary openings in the body-wall.

Exercise 10. Draw a diagram representing an ideal cross section in the region of the stomach-intestine ; show the stomach-intestine with its mesenteries, the blood vessels, the nerves, and the muscles.

ANNELIDA

AN OLIGOCHAETOUS ANNELID. AN EARTHWORM

The earthworm is, to most people, the most familiar annelid. It is distributed over the entire earth, the United States containing many species. The animal is nocturnal in its habits. It lives in long burrows in the ground, in which it lies during the day and the inclement seasons of the year. Its food consists of leaves and other vegetable substances and also of the organic matter contained in the soil which passes through its alimentary canal.

Study the animal first alive, but have one also at hand which has been killed in weak alcohol. Notice its color, or rather lack of color. How is this correlated with its underground life? Note its cylindrical, elongated body, the very small head, and the absence of appendages. Note also the absence of a hard shell, the external integumentary covering being the glistening cuticula which has not been stiffened by the presence of calcareous salts. As the animal lacks appendages, locomotion is accomplished by means of body-movements. Study its method of locomotion. The animal will be observed successively to elongate and to shorten its body, which, of course, would be impossible if it were covered by a hard shell. Notice that along the ventral and the lateral surfaces are several rows of minute bristles, the **setæ**; they aid in locomotion and are under the control of muscles. Determine, by passing the animal through the fingers and with the aid of a hand lens, how many rows there are and their relation to the segments. Determine also whether the setæ at the forward end of the body project in the same direction as those at the hinder end. Observe carefully the importance of the setæ in locomotion.

The animal is segmented externally, *i.e.*, it is made up of a number of **somites** or **metameres**, like the crustacean body. Count the somites, beginning with the segment just back of the mouth, which is the first somite. Notice the equivalence of the somites; they are apparently all very nearly alike. This lack of specialization is always a primitive character in a segmented animal and is in sharp contrast to the condition of the somites in most arthropods. Among the arthropods studied, which most nearly resemble the earthworm in this particular?

Notice the moist, slimy surface. Moisture is necessary to the animal's existence; this accounts largely for its nocturnal habits. Notice also the red blood vessels through the semi-transparent body-wall. What movement of the blood can you detect? What are the differences between the dorsal and the ventral surfaces? Notice the difference between the anterior and the posterior ends. The forward end is the older; the animal grows in length by adding new somites to the hinder end, but the number of somites is practically complete when the young worm emerges from the cocoon. Notice the ventral position of the **mouth** and the terminal position of the **anus**; note also the thickened ring around the body not far from the forward end. This is the **clitellum**; its function will be explained in speaking of the reproductive organs. The animal is without organs of special sense; numerous minute tactile sense organs which are sensitive to light and other stimuli are, however, present. These are distributed along the body but are especially abundant toward its anterior end.

Exercise 1. Make a sketch on a scale of 3 of the ventral aspect of the forward end of the animal back to the posterior border of the clitellum. Indicate the somites and number them.

The body of the animal may be divided into two portions, the **prosoma** or the primitive head, and the **metasoma** or the primitive segmented trunk. The prosoma is further subdivided into the

prostomium, the median dorsal projection overhanging the mouth, and the indistinct **metastomium**, which contains the mouth, and is marked off from the prostomium by fine transverse lines. What somites are included in the clitellum? On the fifteenth somite a pair of prominent transverse slits will be seen. They are the external openings of the **vasa deferentia** or sperm-ducts. On the fourteenth somite look with the hand lens for the two minute openings of the **oviducts**. They are difficult or impossible to see, except during the reproductive period of the animal. Between the ninth and the tenth and the tenth and the eleventh somites are the two pairs of minute openings of the **spermathecæ**, which are also visible only during the pairing season. In each somite, except the first three or four and the last one, is a pair of kidney tubules, called **nephridia**, which open through the body-wall to the exterior by minute pores on either the ventral or the lateral side near the anterior border of the somite. The ventral integument of a number of somites between the seventh and nineteenth is often swollen by the presence of the so-called **capsulogenous glands**. Carefully label in your sketch all of these organs which you have observed.

Exercise 2. Make a similar sketch of the ventral view of the last four somites on a scale of 3.

Internal anatomy. Pin a large worm, that has been killed, firmly to the wax of the dissecting pan by a strong pin at each end; then make an incision with fine, sharp scissors through the integument in the mid-dorsal line from the forward end of the animal to a point back of the clitellum, taking great care not to cut the viscera lying beneath. It will be noticed that the body-cavity is divided into compartments, corresponding to the somites, by transverse partitions, which are called **septa**. Holding the cut edge of the body-wall with the forceps, cut the septa where they join it, and then spread out and pin down the body-wall, using many pins on each side.

Observe first the large **alimentary canal** which passes straight through the animal; also several pairs of conspicuous white bodies a short distance from the anterior end, which are the **sperm-sacs**. If the specimen has been freshly killed, the red blood vessels will also be seen. Study and identify in detail the following systems of organs:

The circulatory system. The earthworm has two circulatory fluids, a red one and a colorless one. The latter consists of a plasma in which float amœboid blood cells. It is present only in the body-cavity and circulates throughout the body, being driven by the movements of the animal from one somite to another through small openings in the septa; it will, of course, not be visible in a dissection. The red blood consists of a red plasma in which float colorless blood cells and it circulates in a system of closed blood-tubes. The most important of these blood vessels are five longitudinal and numerous circular vessels. Observe the **dorsal longitudinal vessel** in the median line, above the alimentary canal. It is contractile and propels the blood towards the head. Push aside the intestine and observe just beneath it the **ventral vessel**, which runs parallel to the dorsal one. Notice that these vessels break into small branches at their anterior ends. The other three longitudinal vessels are very small and not easily seen except in microscopic sections. They lie one beneath and the other two to the right and left of the nerve cord in the mid-ventral line.

The circular or **commissural blood vessels** connect the dorsal and the ventral vessels and have a paired and segmental arrangement. They are not all of equal size. Observe several large pairs near the forward end of the animal which pass directly between the dorsal and the ventral vessels. They are, like the dorsal vessel, contractile and are sometimes called the **hearts**. In which somites are they? Find the **commissural vessels** posterior to them. These are much smaller and do not, in most cases, pass directly between the longitudinal vessels, but break into capillaries between them.

The digestive system. The **pharynx** is an oval, muscular pouch occupying somites 2 to 6; radiating muscle fibers join it with the body-wall. The **œsophagus** is a slender tube occupying somites 7 to 14 and running between the conspicuous sperm-sacs. Press aside these sacs and notice beneath them three pairs of white glands; these are lateral diverticula of the œsophagus and contain calcareous crystals. The **crop** is a thin-walled dilation of the œsophagus which lies in somites 15 and 16. The **gizzard** is a muscular, thick-walled chamber of the same size as the crop and lying in somites 17 to 19. The **stomach-intestine** is a large tube with lateral segmental pouches, which passes to the hinder end of the body; covering the surface of the stomach-intestine is a loose mass of yellowish brown cells, the **chloragogue cells,** whose function is probably excretory.

Exercise 3. Make a drawing of the opened animal on a scale of 3, showing the segmentation and representing the organs above described in their proper somites; label all carefully.

Sever the alimentary tract just back of the pharynx and remove it from the body.

The reproductive system. The earthworm is hermaphroditic and possesses the following genital organs:

The male organs. 1. The **sperm-sacs** have already been noticed. They are large, white, irregularly lobed sacs occupying somites 9 to 13; they vary in size with the sexual condition of the animal, being largest during periods of sexual activity. 2. **The testes.** Two pairs of these organs are present, which lie beneath the sperm-sacs in the tenth and eleventh somites ; they are very minute objects and will be seen with difficulty, if at all. Push aside the sperm-sacs and look for them with the aid of a hand lens. 3. **The sperm-ducts.** These are slender tubes which begin with two pairs of funnel-shaped openings just posterior to the two pairs of testes and in the same somites with them. At the hinder margin of the twelfth somite the two tubes on each side unite to form a single one, and the pair of tubes thus formed run back to the fifteenth somite, where they open through the conspicuous transverse slits already noticed, to the exterior. Look first for the posterior portion of these tubes and trace them forward. The spermatozoa pass from the testes, where they but partially develop, into the sperm-sacs in which their development is completed and where they are grouped together in balls. From here they pass, during pairing, into the sperm-ducts, and out of the animal through the slit-like openings in the fifteenth somite.

The female organs. 1. The spermathecæ. These are two pairs of spherical, white sacs beneath the sperm-sacs in the ninth and tenth somites; they are easily seen. **2. The ovaries.** These are a pair of extremely small organs lying near the median line and attached to the anterior septum of the thirteenth somite near the ventral body-wall; they will hardly be seen. **3. The oviducts.** These are two minute, funnel-shaped tubes which extend from immediately behind the ovaries through the septum to the external opening in the next somite ; they will also hardly be seen.

Earthworms meet and pair in the night time during the months of May and June. Two animals place themselves alongside of each other in such a way that the spermathecæ of each come opposite the openings of the sperm-ducts of the other. The spermathecæ of each are then filled with spermatozoa from the other animal. The worms then separate. Sometime later the clitellum secretes a viscid fluid which hardens and forms a tough cylindrical membrane around the body. The worm then squirms backward, causing this membrane to pass forward toward its head. As the membrane passes the fourteenth somite, eggs are poured from the oviducts into the viscous mass which is held between it and the body, and at the tenth and eleventh somites spermatozoa pass in from the spermathecæ which at once fertilize the eggs. The cylindrical membrane then passes completely off the worm and its two ends close together. It forms thus a yellowish, spindle-shaped capsule about as large as a small pea, and is called the **cocoon**. In it the young animals are born.

Excretory organs. These are the kidneys of the animal. They consist of a pair of coiled tubules, called **nephridia**, which lie near the lateral and ventral wall of the body-cavity in each somite, except the first three or four and the last one. Each nephridium has two openings, a funnel-shaped, ciliated opening into the body-cavity, called the **nephrostome**, and one through the body-wall to the outside. The former in each case is attached to the anterior side of a septum. The tube passes backwards through the septum to the next somite, in which the greater portion of it lies, and through the wall of which it communicates with the outside. The distal, middle, and proximal portions of the tube differ from one another. The **distal portion** (that next to the nephrostome) is very slender, the **middle portion** is much thicker and has glandular walls, and the **proximal portion** is a dilated tube which probably acts as a **urinary bladder**.

Notice the four slight projections in the body-cavity on the ventral side of each somite. These are the **setigerous glands**; they secrete the setæ.

Exercise 4. Make a sketch of somites 8 to 20, representing diagrammatically the reproductive organs and two or three pairs of nephridia lying in their proper somites, and label all.

Crush the sperm-sacs of a fresh worm, that has not been in alcohol, mount some of the milky fluid in it, and examine it under a compound microscope. Notice the **sperm-spheres** and **spermatozoa.**

Exercise 5. Draw a sperm-sphere and a spermatozoan.

With a sharp knife or curved scissors carefully remove a nephridium from the animal's body. Mount it on a slide and examine it under a microscope.

Exercise 6. Draw it and label its three divisions.

The **nervous system** is essentially similar to that in arthropods. Remove the sperm-sacs and observe the **nerve cord** as it lies in the mid-ventral line. Note the slight swellings, which are the **segmental ganglia.** Trace the nerve cord forward to the region of the mouth, where it encircles the forward end of the pharynx and joins the small **brain.** Observe the two ganglia of which the brain is composed. Remove the forward portion of the nervous system, together with the brain, from the body. Mount it on a slide and examine it under a microscope. Note the double nature of the nerve cord and of the ganglia. What does this signify as to the primitive condition of the system in the ancestors of the earthworm? Note accurately the lateral branches that leave the cord; also the shape and branches of the brain.

Exercise 7. Draw the nervous system on a large scale, accurately representing all the details.

Study of a **cross section.** This is instructive because it shows the relations of the organs to one another in their natural positions and also illustrates their finer structure. A properly stained and mounted cross section of any portion of the body will serve for this study.

Observe first the **integument**; it is made up of the **cuticula** on the outside and the cellular **hypodermis** beneath it. The latter is composed, in most parts of the body, of a single layer of cells and it secretes the cuticula. Note the numerous single-celled **glands** in the hypodermis. If the section passes through a **seta,** notice its method of attachment and its muscles. Beneath the integument are the body-muscles. Of these there are two systems, the **circular** and the **longitudinal muscles**. The former are a narrow band just beneath the hypodermis. The latter are much more extensive and project into the body-cavity; they are arranged in groups and will be seen of course in cross section. Near the center of the body-cavity note the large **alimentary canal**. If the section be in the region of the stomach-intestine, note the longitudinal fold in the dorsal intestinal wall which very largely increases its surface. Observe the structure of the alimentary canal; its cavity or lumen is bounded by a thick mucous membrane consisting of a single layer of very long, slender cells, around which are two muscle layers, an inner **circular** and an outer, very thin, **longitudinal** layer. Surrounding the muscle layers and also forming a thick fold over the dorsal and lateral intestinal surfaces are the pear-shaped **chloragogue cells**. Observe the **dorsal** and the **ventral blood vessels**, and also the **commissural blood vessels**, if any are in the section. Study carefully the **nervous system**. Note the **muscular sheath** which surrounds the nerve cord, and imbedded in it the **subneural** and the two **lateroneural blood vessels**. Note the double nature of the nerve. Note the large pear-shaped **nerve cells**, and the **nerve fibers**, also the three large bodies in the dorsal portion of the ganglion. These latter are called the **giant fibers**. Do lateral nerves join the ganglion? If so, trace their fibers into it. Also trace their fibers away from the ganglion and see where they go. Examine carefully the **peritoneum**. This is a layer of cells which lines the body-cavity and bounds all the organs in it.

Exercise 8. Draw the cross section and carefully label all the organs.

CHAPTER III

PLATHELMINTHES

TURBELLARIA

A PLANARIAN WORM

Planarian worms are very common animals in freshwater streams and ponds as well as in the sea; they may be found on the underside of stones or on aquatic vegetation. They are flat, elongated, very soft and contractile animals, brownish or yellowish in color, and usually half an inch or less in length; at the forward, broader end, on the dorsal surface, are two black eyes; the hinder end is pointed. A variety of forms is found, some of which are very minute and are without an intestine or have a straight, tubular intestine, while others are much larger and have a branched intestine. The latter include most of the commoner turbellarians and those for which these directions have been prepared.

Study the live animal under a dissecting microscope. Note the gliding motion with which it moves. This is accomplished partly by the action of the cilia which cover its surface and partly by muscular contraction. In the middle of the ventral surface are the **mouth** and the protrusile **proboscis**. Mount the animal on a slide in water beneath a thick coverglass and observe the action of the cilia under a compound microscope.

Exercise 1. Draw an outline of the animal on a large scale, with the eyes and proboscis, and indicate its anterior and posterior ends.

Study an animal which is under the pressure of a large cover-glass, and make out as many of the following organs as possible, using often reflected instead of direct light:

The digestive system. The digestive canal is usually easily seen. The **mouth** is a circular opening near the center of the ventral surface; it leads into the **pharynx**, a cylindrical organ with thick, muscular walls, which can be thrust out of the mouth as a proboscis. At the base of the pharynx the **intestine** divides into three trunks, one of which passes forward, and the other two backward to the extremities of the animal's body. Each of these trunks gives off lateral branches which are themselves often branched. There is no anus.

Exercise 2. Draw an outline of the animal and place in it the digestive system in detail.

The reproductive system. Planarians are hermaphroditic; the sexual organs are complicated in structure and arrangement and difficult to observe in a live specimen. Near the lateral edges of the body will be seen, among the ends of the lateral intestinal branches, two sets of lobed organs. Of these the larger are the **yolk glands**, which connect with the **oviducts**; the smaller and less apparent ones are the rounded **testes**. Just back of the mouth is the **uterus**, which is often to be recognized by the spherical eggs it may contain; it passes back to a sac called the **genital cloaca**. The **ovaries** are a pair of spherical bodies in the anterior part of the body, and from them a pair of **oviducts** extends to the hinder part of the body, receiving the lateral yolk glands on their course. Leading from the **testes** are the **vasa efferentia**, very delicate tubes, which pass to the conspicuous **vasa deferentia**. There is a pair of the latter organs, one on each side of the mouth and pharynx, and they extend to the hinder part of the animal, where they unite to form the muscular **cirrus**, which opens into the genital cloaca. The two **oviducts** also fuse at their hinder ends, and the median duct thus formed opens into the **genital cloaca**.   This structure, which thus receives all the ducts from the genital glands, communicates with the outside through the **genital pore**, a median, ventral opening in the hinder part of the body.

Exercise 3. Draw a large outline of the animal and place in it as much of the reproductive system as you have observed.

The nervous system.   Beneath the eye-spots will be seen the opaque **brain**, a large nervous mass consisting of a pair of minor masses united by a broad commissure.   From its anterior and lateral sides numerous **sensory nerves** pass to the anterior body-surface, which render this extremity a highly sensitive tactile organ.   From its posterior side a pair of large **longitudinal nerve cords** passes to the hinder end of the body, being united at intervals by **transverse nerves**.

The excretory system.   This consists of a system of minute tubes which extend throughout the body and collect the excrete matters from the tissues.   There are two main **longitudinal tubes** extending the length of the body, which open to the outside through minute pores on the dorsal surface of the animal. These tubes are not straight but coiled and give off numerous branches, at the termination of each of which is a peculiar cell with a vibratory process at its base called a **flame cell**; they are joined by a **transverse tube** at the anterior end of the animal. Portions of the excretory system can often be seen in the compressed animal, where they appear as fine lines.

Exercise 4. Draw an outline of the animal and place in it as much of the nervous and the excretory system as you have observed.

No special respiratory system is present in the Turbellaria, the ciliated outer surface of the body performing this function. A circulatory system and a blood fluid are also wanting.   The branching of the digestive and excretory systems is correlated with this feature. Can you explain how? As in other flat-worms, the turbellarians possess no body-cavity, the primitive body-cavity being filled secondarily by a peculiar vesicular connective tissue called **parenchyma**. The **muscular system** consists of a layer of strong circular and longitudinal muscles just beneath the surface of the body and of oblique muscles passing through the parenchyma.

CESTODA

A TAPEWORM

Tapeworms are common parasites in the intestines of vertebrate animals. The human tapeworm, *Taenia saginata*, may often be obtained from physicians. If it be alive when obtained it should be placed in a normal salt solution (a 0.75% solution), in which it will keep alive for several hours, and may then be studied. If it be dead it should be preserved in alcohol or formalin. *Taenia serrata*, a tapeworm of the dog, and *Taenia crassicollis*, which lives in the cat, are both common animals and are convenient forms for study. The intestines of adult cats or dogs should be slit open and the worms taken out and placed alive in a normal salt solution. They are white, band-like objects, six inches or more in length, which are attached by one end to the wall of the intestine. Care should be taken in separating them from the intestinal wall not to tear them.

Study the movements and general form of the animals as they lie in the salt solution. The worm will be seen to be made up of a large number of segments, and to bear at the smaller end a small rounded knob. The segments are called **proglottids**, and the rounded knob, the **scolex**. The body of the animal is not made up of body-divisions which we can call head and trunk. The scolex, however, may be held to represent its anterior end, the proglottids having arisen from it by a process of terminal growth. The scolex is thus the oldest part of the animal's body; it, in fact, constitutes the entire parasite when it first arrives in the intestine of the host (as the animal is called in which a parasite lives), the proglottids only then beginning to grow. The youngest proglottids are those nearest the scolex; those at the opposite end of the body are the oldest and hence the largest. Count the proglottids. The animal attaches itself to the wall of the intestine by means of its scolex, which is provided for this purpose with four **suckers** and usually two rows of chitinous **hooks**; the scolex of *Taenia saginata* lacks the hooks. Thus attached, it lies immersed in the digestive fluids of its host and absorbs through the outer surface of its body the nutriment it needs. It is without a digestive system.

Exercise 1. Draw an outline of the animal on a scale of 4 or 5, taking care to represent the number of proglottids accurately.

The scolex. Cut off the scolex, mount it on a slide in glycerine or water, and examine it under the microscope. Notice the fine **excretory canals** which occur in every part of it. Can you determine their arrangement? Note the numerous minute **calcareous bodies**.

Exercise 2. Draw the scolex on a scale of 10. Represent accurately the suckers and the number and position of the hooks, if these are present.

Exercise 3. Draw a single hook highly magnified.

The proglottids. Each proglottid is composed mainly of reproductive organs and circular, longitudinal, and oblique muscle fibers imbedded in a spongy tissue called **parenchyma**. The parenchyma fills the entire primitive body-cavity, which is thus absent in this animal. Each proglottid contains a complete set of both **male** and **female genital organs**. These are immature in the youngest and smallest proglottids; in those at about a third of the distance from the anterior end of the body they are mature; in the largest proglottids, those at the posterior end of the body, the **uterus** is so distended with eggs that most of the other genital organs are obliterated and do not appear. A pair of longitudinal **excretory canals** passes from one end of the worm to the other, running near to and parallel with each lateral edge; in each proglottid, also, are one or two **transverse canals.** One or more pairs of **longitudinal nerves** run parallel with and very near the excretory canals, which are also joined in each proglottid by a **ring commissure.**

Cut off two or three proglottids from the forward end of the body, two or three from about a third of the distance from the anterior end, and two or three from the posterior end, and soak them all first for a short time in a solution of caustic potash and then in one of equal parts of glycerine and water.

Place the proglottids from the hinder end of the body in dilute glycerine between two glass slides, press them gently so as to squeeze them as thin as possible without crushing them, and study them under the microscope. In the middle of one of the lateral edges of the proglottid notice a slight projection containing a depression. This is the **genital papilla,** and the depression is the **genital cloaca.** Two canals will be seen running from the genital cloaca toward the center of the proglottid. These are the **vas deferens** and the **vagina,** the former being the larger of the two. The greater part of the proglottid will be seen to be occupied by a much-branched organ filled with a granular substance. This is the **uterus,** and the granules are eggs. Near each lateral edge of the proglottid will be seen a straight band. These two bands are the dorsal **excretory canals**; find the **transverse canal** near the hinder end of the proglottid. Between each canal and the edge of the proglottid will be seen a delicate line running parallel with the canal. These are the main **nerves**; they are joined by transverse nerves.

Exercise 4. Draw the proglottid, showing accurately all of these features that you have observed.

Study in the same way the mature proglottids. Find the **uterus.** It is here a straight, narrow tube in the middle of the proglottid, and is not yet distended with eggs. Near the center and toward the posterior end of the proglottid will be seen an irregular mass of organs. These are the paired **ovaries,** two large, round bodies, one on each side of the uterus; the median **yolk gland,** which is below the end of the uterus, near the posterior margin of the proglottid; the **shell gland,** between the yolk gland and the uterus. From the shell gland the **vagina** and **vas deferens** proceed to the **genital cloaca,** the former being the smaller and more posterior of the two. Scattered throughout the proglottid are numerous small round bodies, the **testes,** which are joined with the vas deferens by numerous minute **vasa efferentia.** Find the **excretory canals** and the **longitudinal nerves.**

Exercise 5. Draw the proglottid, showing all of these features you have observed; carefully label all.

Study in the same way the immature proglottids from the forward end of the body. Find as many of the organs mentioned as are present.

Exercise 6. Draw the immature proglottid.

The tapeworm may fertilize itself or be fertilized by another individual, and where self-fertilization takes place one proglottid of the animal may fertilize another, or a single proglottid may fertilize itself. The ova from the ovaries, on being fertilized, pass at once into the uterus. The ripe proglottids, which are filled with eggs in which the embryo has already begun to develop, break off from the hinder end of the worm and pass out of the body of the host. They then break open or are crushed, and their eggs are scattered on all sides.

The encysted tapeworm. The adult worm alone is found in the intestine. The eggs, in order to develop, must pass out of the host and fall upon something which will afterward be eaten by another animal, called the **intermediate host**, which is itself preyed upon by the host. After being thus transferred to the stomach of the intermediate host there hatches from each egg a minute spherical embryo, called the **six-hooked embryo**, which is provided with three pairs of hook-like organs of locomotion. This embryo works its way through the wall of the intestine of the animal and migrates finally to some one of the internal organs, where it lodges and grows into a cyst-like larval form, called the **cysticercus**. Within the cysticercus is a fully developed scolex, but turned wrong side out. If, now, the intermediate host be eaten by the host, the scolex turns right side out, passes into the intestine of the latter, attaches itself to the intestinal wall, and grows into an adult tapeworm.

The intermediate host of *Taenia saginata* is the beef, in the muscles of which the cysticercus will be found, if present. That of *Taenia serrata* is the rabbit and that of *Taenia crassicollis* is the mouse; in the former animal the cysticerci are imbedded in the peritoneum or the liver, and in the latter, in the liver. Open the body-cavity of either of these latter animals by a median ventral incision and look for cysticerci. They are large, whitish bodies and are easily detected if present. When a cysticercus is found, it should be carefully dissected out, its outer wall slit and the scolex exposed to view. Mount it on a slide in dilute glycerine and study it.

Exercise 7. Draw a view of the scolex in its cyst.

CHAPTER IV

BRYOZOA (POLYZOA)

ECTOPROCTA

AN ECTOPROCT BRYOZOAN (*Bugula turrita*)

Bugula turrita is a marine colonial bryozoan which is very common in the shallow waters along our coast. The colonies are sessile and are attached to rocks, seaweed, and other objects. The animals are very small and must be studied with the aid of a microscope.

Study a large piece of a colony (alive if possible) and notice the spiral arrangement of the branches. A branch is made up of a double row of elongated partitions or chambers, each of which is called a **zoœcium**. Each zoœcium represents a separate individual of the colony ; within its walls are the soft parts of the animal which are called collectively the **polypide**. The individual bryozoan is thus made up of two distinct parts, the zoœcium and the polypide, the former constituting the chitinous outer wall of the animal, the latter comprising its viscera and the tentacles. At its upper or distal end the zoœcium has a large opening through which the forward end of the polypide can be protruded and into which it withdraws itself when alarmed. The **cuticula** which forms the zoœcium is rendered hard by the presence of carbonate of lime ; it is thus much more enduring than the remainder of the animal, and after death the empty zoœcium may persist long after all the softer parts have disappeared. Look for empty zoœcia in your specimen.

The different individuals of a colony have arisen by a process of budding from the individuals below them in the colony. The oldest individuals are thus those nearest the base of the colony, the basal one being the progenitor of the entire colony. This is also the only individual which has not come into existence by a process of budding; it began its life as a free-swimming larva which was hatched from an egg.

The zoœcium.   Mount a small portion of the colony containing two or three branches on a slide under a cover-glass.

Exercise 1. Draw a large and accurate outline of the zoœcia, leaving out the polypides. Observe very carefully the boundaries of the zoœcia and their relations to one another.

The polypide.   Study a number of polypides, both retracted and extended. The forward end of the polypide consists of a circular ridge, called the **lophophore,** which bears a row of long ciliated **tentacles.** In the midst of the circle is the **mouth.** The tentacles are very vibratile and serve as respiratory as well as prehensile organs. It will be seen that the lophophore can be entirely withdrawn within the zoœcium.

The digestive system.   The **mouth** opens into the **pharynx,** which leads into the **œsophagus.** This opens into a large sac-like **stomach,** the lower portion of which is lengthened into a long pouch. From the upper end of the stomach, near the base of the œsophagus, the short **intestine** leaves it and passes to the thick-walled **rectum.** This leads to the **anus,** which is situated just outside the lophophore near the mouth. The digestive tract has thus the shape of the letter V, the point of which is formed by the stomach pouch. Passing from the stomach pouch to the lower end of the body is a broad mesenteric strand called the **funiculus.** In order to study the digestive tract satisfactorily, a polypide should be found in which both arms of the V come into view.

The muscular system. The retractor muscles, whose function it is to draw in the lophophore, may, in favorable specimens, be seen as delicate strands which pass from the wall of the zoœcium to the pharynx.

The **nervous system** has not yet been observed in Bugula, but in nearly allied Bryozoa it consists of a single ganglion between the mouth and the anus. From it nerves radiate to the tentacles and other organs. There are no organs of special sense.

The reproductive organs. The animals are hermaphroditic. Ova develop from the peritoneal lining of the spacious body-cavity and will be seen, when present, lying near the stomach pouch. Spermatozoa develop from the funiculus and, when present, form a mass about that organ.

There are two methods of reproduction, the sexual, in which the new individual develops from the fertilized egg, and the asexual, in which the new individual arises by budding. As already stated, the entire colony, with the exception of its oldest member, has developed in the latter way.

Exercise 2. Draw an extended individual in which the entire digestive tract can be seen and label all the organs observed.

There are no special respiratory or excretory organs; the entire outer surface of the body performs these functions. The **circulatory system** is represented by the colorless blood fluid alone. There are no circulatory vessels, the blood being contained in the body-cavity.

Avicularia and oœcia. These are peculiar structures, found in connection with the zoœcia, which are morphologically equivalent to distinct individuals. An **avicularium** is a small structure, like a bird's head in shape (hence the name), which may be found attached to the wall of some of the zoœcia near the opening. It has a movable lower jaw which can be opened and shut by two sets of muscles. Its function is to seize and hold small animals. These soon die in its grasp, and their disintegrated remains are swept into the mouth by the ciliated tentacles. An **oœcium** is a disc-like structure which, in some parts of the colony, lies in front of a zoœcium. It serves as an egg-capsule in which the embryo develops. A single embryo will be found in each.

Exercise 3. Find a zoœcium with an avicularium attached and draw them.

Exercise 4. Find a zoœcium with an oœcium attached and draw them.

CHAPTER V

MOLLUSCA

PELECYPODA

A FRESHWATER MUSSEL (*Anodonta* or *Unio*)

These animals are common in most parts of the country; they inhabit the sandy bottoms of freshwater streams and lakes.

Study first the live animal, if possible. Its body is unsegmented and entirely enclosed in a bilateral, bivalve shell, which is the cuticula of the animal richly charged with calcareous salts. The two valves of the shell cover the right and left sides of the animal and are joined together on its dorsal side by the dark-colored **hinge ligament**, while their ventral edges are open; the animal is thus very much compressed laterally. The anterior end of the animal is more rounded and less elongated than the posterior end. Which is the right-hand valve? The elevation on each valve near the hinge towards the forward end is called the **umbo**. It is the oldest portion of the shell; from it as a beginning point the shell has grown in size by additions to its ventral edge. Note the parallel lines of growth. The ventral edges of the shell are thus the youngest portions of them.

Exercise 1. Make a drawing of the right-hand valve, indicating the anterior, posterior, dorsal, and ventral aspects, and showing the lines of growth.

Exercise 2. Make a drawing of the dorsal aspect of the animal.

Kill the animal by immersing it for a few minutes in hot water (70° C.). As the shell is kept closed by the contraction of the two muscles which pass between its valves, it will gape open as soon as the animal is dead and the muscles are relaxed. It is the elasticity of the hinge ligament which causes it to open.[1]

Examine the animal as it lies in the shell. It will be seen that the inner surface of each valve is covered with a soft, slimy membrane, the lower edge of which is parallel to the edge of the shell. This is the **mantle**; it is a double fold of the dorsal integument of the body, one side of the body being covered by either fold. The mantle is the **matrix** of the shell, *i.e.*, it secretes the shell. Its lower edge is provided with muscle fibers and can be extended beyond the edge of the shell; it also possesses sensory functions; in some pelecypods eyes are situated along the mantle's edge.

Observe the large, soft **visceral mass** hanging between the two lobes of the mantle; it contains the viscera of the animal. On the lower side of the visceral mass, *i.e.*, toward the gape of the shell, is the muscular **foot**, which can be extended below the edge of the shell and is the organ of locomotion. Observe the two leaf-like **gills** on each side of the visceral mass and foot; also the two large **adductor muscles**, one in front of and the other behind the visceral mass, which pass from one valve of the shell to the other and serve to close them.

Pass a knife between the mantle and the left shell and separate them from each other. Cut the two muscles close to the shell; cut the elastic hinge ligament and remove the left shell.

Study the inner surface of the shell. Note the two large scars marking the surfaces of attachment of the adductor muscles;

---

[1] The shell may also be opened by inserting some sharp, wedge-shaped instrument between its valves. The valves are thus pressed apart far enough to admit the blade of a scalpel, by means of which the adductor muscles should be cut close to the left valve of the shell. The hinge ligament should then be cut and the left valve be removed.

just above each is the scar of a much smaller muscle, the **retractor** of the foot, and just posterior to the anterior adductor muscle is the scar of the **protractor** muscle of the foot. Note the broad line which joins the scars, running parallel with the edge of the shell. This is the **pallial line**; it is formed by the insertion in the shell of the delicate muscle fibers at the edge of the mantle. Do you find **hinge teeth** in the shell just beneath the hinge ligament? **Unio** has such teeth; **Anodonta** is without them and is also characterized by the thinness of its shell.

Exercise 3. Draw a view of the inner surface of the shell.

Break the shell and examine the broken edge with a hand lens. Study the structure of the shell. It is composed of three layers — the inner **mother-of-pearl layer**, which is secreted by the entire surface of the mantle, the **prismatic layer**, and the **organic layer** or **periostracum** on the outside. The two latter layers are secreted by the edge of the mantle; the periostracum is very thin and gives the color to the shell. Place a piece of the shell in a solution of hydrochloric acid; note the effervescence which results; also that an organic remnant, even of the two inner layers, is left.

Exercise 4. Draw a view of the broken edge of the shell on a scale of 5. Show the prisms of the prismatic layer.

Place the animal in water and study it as it lies in the right shell.[1] The two halves of the mantle will be seen to envelop entirely the visceral mass and the foot. Over the entire dorsal portion of the visceral mass the mantle is fused with it and cannot be separated, but the lateral and ventral portions of the mantle lobes hang free, enclosing an extensive space, which is called the **mantle cavity**. In it, on each side of the visceral mass,

---

[1] For the study of the soft parts of the mussel it is well to have at hand also a specimen which has been deprived of both valves of the shell.

lie the two leaf-like **gills**. In front of the gills are two pairs of triangular flaps, the **oral palps**, between which, in the median line and just back of the anterior adductor muscle, lies the **mouth**. Find it.

Trace the irregular line of attachment of the mantle with the visceral mass; it follows the base of the gills and of the oral palps and passes beneath both adductor muscles. Observe the edges of the mantle and note that at the hinder end of the animal they are darkly pigmented, and the middle point of the pigmented line is joined with the base of the gills by a short **septum**. This septum divides the posterior portion of the mantle cavity into a **dorsal** and a **ventral chamber**. The latter is the very large **branchial chamber** which contains the gills; the former is the very small **cloacal chamber**. The pigmented edges of the mantle are at this place modified to form, when the edges of the two sides of the mantle are applied to each other, two short tubular openings, which place these two chambers in communication with the outside water and are called the **siphons**. The ventral siphon is called the **branchial** or **incurrent siphon**; through it water streams into the branchial chamber bearing food and air for respiration. The dorsal siphon is called the **excurrent** or **cloacal siphon**, and through it water passes outward charged with fæcal matter from the alimentary tract and carbon dioxide of respiration. Note the sense tentacles on the branchial siphon.

Probe the dorsal siphon. Carefully remove the left lobe of the mantle after cutting it with fine scissors along its line of attachment with the visceral mass.

Through the transparent body-wall observe the organs in the dorsal portion of the visceral mass. Just back of the anterior adductor muscle is the **liver**, which can often be recognized by its greenish color, and back of which is the dark-colored **Keber's organ**. Between the hinge ligament and the base of the gills lies the **heart** in its transparent **pericardium**, and beneath it is the dark-colored **kidney**. The **rectum** may be seen passing through the pericardium and the heart, then extending above the posterior adductor muscle to the **cloaca**, where it ends with the **anus**. Cut open the cloacal chamber by a slit in the side of its siphon. Find the hinder end of the rectum and the anus. Note just beneath the muscle a canal which accompanies the base of the gill forward. This is the **suprabranchial passage** of the outer gill; it runs posteriorly to the cloacal chamber. Blow into it with a blow-pipe, also probe it from behind.

Exercise 5. Draw a semidiagrammatic view of the animal lying in the right-hand valve of the shell, representing the organs above mentioned. Carefully label all.

The respiratory system. The gills have already been noticed. The two gills on each side of the visceral mass are, by way of origin, but a single organ, which is called the **ctenidium**. The mussel is thus provided with a single pair of ctenidia, which are homologous to those of the squid and of snails. Each gill consists of a pair of plates or lamellæ united at their lower edges and open above, and further joined by vertical or dorso-ventral cross-partitions, the **interlamellar partitions**. The space between the lamellæ is thus divided into parallel, vertical chambers, the **water-tubes**, which run from the bottom to the top of the gill and open above into the **suprabranchial passage**. One of these passages runs along the base of each gill, as a wide canal, to the cloacal chamber. We have already observed the suprabranchial passage of the outer gill. In order to observe that of the inner gill, lift up both gills; the inner lamella of the inner gill will, in most species of mussels, be seen not to be united with the wall of the visceral mass along the hinder portion of the foot, but to have a free edge. The long slit-like opening thus presented leads into the inner suprabranchial passage. Probe it backward to the cloacal chamber. Probe it also from the hinder end forward and notice that back of the visceral mass the two inner suprabranchial passages, *i.e.*, those belonging to the inner gills of the right and left sides, coalesce, forming a single passage.

Study the finer structure of the gills. Place a gill on a glass slide in a little water and with forceps and a knife carefully separate the lamellæ. Mount a piece of a lamella in water and study it under a compound microscope. Note the vertical inter-lamellar partitions. Observe that the lamella is a delicate lattice work made up of ridges, the **gill-filaments**, running vertically and thus parallel with the interlamellar partitions, and of cross-ridges, the **interfilamentary connections**, which run between and connect the vertical filaments. The apertures in the lattice work place the water-tubes in communication with the water in the branchial chamber. The gill-filaments are provided with cilia, as may be easily seen if the gill be alive, the action of which causes streams of water to pass into the water-tubes. The course of the respiratory water is from the branchial chamber into the water-tubes, through which it passes into the suprabranchial passages and through these into the cloacal chamber.

Exercise 6. Draw a diagram of the respiratory system showing the gills and their relation to the suprabranchial passages. Show the direction of the flow of the respiratory water by means of arrows.

Exercise 7. Draw a diagram showing the minute structure of a lamella.

The circulatory system. With fine scissors and great care cut open the **pericardium** by a slit along its dorsal border. Note the **heart** with the rectum passing through it. The heart consists of three chambers — a median, thick-walled **ventricle** and two lateral **auricles**. These latter are delicate, thin-walled organs, triangular in shape, the base of the triangle lying along the dorsal border of the gills and the apex communicating with the ventricle. If the left auricle has been injured in the dissection, the right one is easily seen by looking across the pericardial space. From the ventricle an anterior and a posterior artery pass to either end of the body. These arteries lie alongside the rectum, to which the anterior one is dorsal and the posterior one is ventral; they are difficult to distinguish from it, except in specimens in which the heart has been injected.

The course of the blood is the following: by the contraction of the heart the blood is sent to all parts of the body; on its return course it is first conveyed, through a system of lacunæ, to the kidneys, and thence to the gills; here it circulates in vessels which run through the interlamellar partitions, the gill-filaments, and the interfilamentary connections, and is oxygenated; it then passes into the auricles.

The **excretory system** consists of a pair of **kidneys**, which are dark-colored organs lying just beneath the pericardium and in front of the posterior adductor muscle. Each kidney consists of two parts, the **kidney proper** and the **ureter**. The former is a dark, thick-walled gland which lies beneath the ureter and communicates with it at its hinder end. The ureter is a thin-walled vessel lying above the kidney proper, with a small external opening on the side of the visceral mass beneath the anterior end of the kidney and near the base of the inner gill. With fine scissors cut off the gills and look for the opening; it may be recognized by its white lips. The kidney also possesses at its anterior end a duct leading into the pericardial cavity. Slit open the ureter and kidney proper in clean water and observe their inner structure.

Exercise 8. Draw a diagram representing the pericardial cavity and the kidney, showing the relation of the two structures to each other. Draw the heart in the pericardial cavity, showing the relation of the auricle to the gills.

The digestive system. Find the **mouth** between the two pairs of palps and place a bristle in it; note the upper and the lower **lips,** which connect the upper and the lower pairs of palps, respectively. The mouth is seen to the greatest advantage in a specimen which has been deprived of both valves of the shell. Trace the rectum from the anus to the place of its entrance into the visceral mass. Carefully remove with forceps and knife the tough, white integument which covers the left side of the visceral mass, taking care not to disturb the organs beneath. The soft cream-colored mass just above the foot is the reproductive gland; the light greenish mass lying just above this is the liver. Imbedded in these masses lies the **alimentary tract,** a narrow delicate tube, which will be injured in the dissection unless the greatest care is taken. Beginning with the mouth, gently scrape away the soft mass which surrounds the alimentary tract, laying it entirely bare. The water in the dissecting pan must be frequently renewed to keep it clear, and great care must be taken not to break the tract. The mouth opens into the short œsophagus, after which the canal dilates to form the **stomach.** The liver communicates with the stomach by several ducts. Back of the stomach is the **intestine,** a narrow tube which runs backward and downward to the hinder end of the visceral mass; it then turns upward and runs forward to a point above the stomach, where it turns downward to the lower side of the visceral mass; it then bends dorsally again and runs to the point where it leaves the visceral mass. Here the **rectum** begins and passes through the heart and above the posterior adductor muscle to the **anus** in the cloacal chamber.

Mussels feed upon minute organisms and organic particles contained in the water. Some of the water in the mantle cavity is drawn by the ciliated oral palps into the mouth and passes through the alimentary tract, where organic substances contained in it are digested and absorbed. The mussel usually lies with its forward end buried in the sand and its hinder end with the siphons projecting into the water.

Exercise 9. Draw a diagrammatic view of the digestive system.

The reproductive system. The sexes are separate. The **reproductive glands** (testis or ovary) are very similar to each other and consist of a pair of cream-colored masses which fill a greater part of the visceral mass. They communicate with the outside through a pair of small openings, one on each side of the visceral mass just below and in front of the opening of the ureter. The openings can often be located by pressing out from them eggs or sperm. The eggs, as soon as laid, pass into the interlamellar space of the outer gills of the mother, where they hatch. The young larvæ are very immature and are called **glochidia**; they leave the mother and attach themselves to the sides of fishes by means of a pair of sharp projections on the ventral edges of the valves of the shell, where they lead a parasitic life. While here they undergo a metamorphosis and finally attain the adult structure, when they detach themselves and drop to the bottom. Look for glochidia in your specimen.

The **nervous system** consists of three pairs of ganglia — the **cerebral ganglia** or **brain**, the **pedal ganglia**, and the **visceral ganglia**, and the nerves proceeding from them; each of the last two pairs is joined with the brain by a pair of nerve-connectives.

First find the visceral ganglia. They are a small grayish or pinkish mass on the ventral surface of the posterior adductor muscle with nerves radiating in all directions. Two of these nerves, the **cerebro-visceral connectives**, will be seen passing forward, one on each side of the visceral mass.

Find next the brain. It consists of a pair of ganglia situated above the mouth, just behind the anterior adductor muscle. The two ganglia are not so close together as those of the visceral pair; they lie on either side of the muscle and are united by a commissure. Each ganglion sends out three large nerves — the

cerebro-visceral connective, which goes to the visceral ganglia, the **cerebro-pedal connective,** which goes to the pedal ganglia, and the **pallial nerve,** which passes to the mantle. Find them.

The pedal ganglia form a nervous mass buried in the foot near its base. Make a shallow longitudinal incision in the bottom of the foot and gently pull the flaps apart; the pinkish mass and the nerves radiating from it will be seen. In contact with it is a sense-organ called the **otocyst.**

Exercise 10. Draw a diagram representing the nervous system.

Make several transverse sections with a razor through the region of the heart of a mussel which has been previously hardened. Identify all the organs which appear.

Exercise 11. Draw a diagram representing a cross section; carefully label all the organs.

PELECYPODA

AN OYSTER

Select a large live oyster in the shell, and if it is dirty wash it thoroughly. The shell is sometimes covered with mud, hydroids, sponges, tube-forming annelids, and other marine animals. The small, round holes made by the yellow boring sponge are often conspicuous.

The two valves of the shell will be seen to be different in shape, one being more or less flattened and the other much deeper and more convex. These two valves cover the right and left sides of the animal's body, the convex valve being on the left and the flattened one on the right side. The oyster is a sessile animal, after it has passed through its youthful migratory period, and is fastened to a rock or shell or other stationary object by its left shell. It thus lies on its left side, while the flat right shell acts as a cover which can be raised to allow the animal to draw in water containing food and air, and closed when danger threatens. The very young oyster is a symmetrical animal which swims about actively in the water. While it is still very small — so small, in fact, that it is barely visible to the naked eye — it settles down and fastens itself to some stationary object and in its subsequent growth accommodates itself more or less to the irregularities of this substratum. This is the reason why the shell is so often rough and irregular in shape.

The smaller end of the shell is the anterior end. The **hinge ligament** is situated here, the elasticity of which keeps the shell open except when it is closed by the contraction of the large adductor muscle. At this end is also the **umbo**, the oldest part of the shell. Note the parallel lines of growth which extend from the umbo to the ventral and posterior sides of the shell. When the anterior, the right, and the left sides of the shell are known, the ventral and posterior sides can be easily determined.

Exercise 1. Make an outline drawing of the right valve, indicating the anterior, posterior, dorsal, and ventral aspects and showing the lines of growth.

Remove the right valve in the following way: Break off the edge of the shell with a hammer, insert the blade of a scalpel and cut the large adductor muscle, which is not far from the edge but nearer the dorsal than the ventral margin. It is important to keep the blade close to the right valve so as not to mutilate the internal organs. Force off the right valve and examine its inner surface.

Exercise 2. Draw the inner surface of the shell, showing the muscle scar with its lines of growth and the hinge ligament, and label the dorsal, ventral, anterior, and posterior sides of it.

Study the animal as it lies in the left valve. Note the soft, shiny **mantle**, which covers the inner surface of the shell and has secreted it. The mantle is a double fold of the integument which extends ventrally from the dorsal side and covers the two lateral sides of the body. Its lower edge is bordered by a fringe of short, pigmented **tentacles** which are the principal sense organs of the animal; it is also provided with muscle fibers which enable it to be slightly extended beyond the edge of the shell.

The most conspicuous organ in the body will be seen to be the large **adductor muscle**. Lying between it and the hinge ligament is the **visceral mass**, containing most of the viscera. Along the ventral side are the four **gills**.

Put the oyster into a pan of water and with fine scissors and forceps remove the right mantle. Just in front of the adductor muscle observe the **pericardium**. Carefully cut it away and see the **heart**, which lies in the pericardial cavity; it will be beating if the animal is still alive. The **ventricle** is dorsal in position and the **auricle** is ventral, lying next to the gills, from which it receives the purified blood. The four gills lie close together, no foot being present to separate the two right-hand from the two left-hand gills. Just in front of the gills, at the front end of the body, are the two pairs of large **oral palps**. The **mouth** is between these palps, two being on each side of it. Find the mouth and note that it lies between an **upper** and an **under lip**, each of which is formed by the union of a pair of palps; *i.e.*, a palp on the right side joins one on the left and forms the upper lip, and the other two palps join to form the under lip.

Oysters feed on minute organisms contained in the water. These are caught in the slime which exudes from the surface of the gills and moved forward by the action of the cilia of the gills and the palps to the mouth.

The **anus** and the **rectum** will be seen on the dorsal side of the adductor muscle.

Exercise 3. Make a drawing of the oyster as it lies in the left shell, representing all the organs above mentioned. Carefully label all.

The digestive tract. This consists of the short œsophagus, the **stomach** and the dark-colored **liver** which surrounds it, and the long **intestine**. The mouth opens directly into the œsophagus, which leads to the stomach. The position of this organ can easily be determined, because it is imbedded in the dark-brown liver. Carefully scrape or cut away the side of the visceral mass and expose the liver; continue the process until the stomach is seen. The intestine extends straight back from the stomach to a position ventral to the adductor muscle and between it and the gills. It then turns on itself and passes straight forward to the dorsal side of the stomach, around the forward and ventral sides of it, and thus back again to the dorsal side of the muscle, where it ends with the anus. Most of it is surrounded by the yellow **reproductive gland**. Lay bare the intestine. This can be done best after the oyster has been hardened for a few days in a 5 per cent. solution of formalin.

Exercise 4. Make a drawing of the digestive tract in an outline of the animal's body.

The remaining systems of organs of the visceral mass will not be studied in this dissection.

The American oyster is a unisexual animal; the common European oyster is hermaphroditic. The reproductive glands, the **ovaries** or **testes**, are a pair of yellowish or whitish organs of irregular form which occupy the larger part of the visceral mass and surround the digestive tract and other organs. The **kidneys** are also a pair of organs of irregular form which, together with a portion of the intestine, occupy the lower and hinder part of the visceral mass, between the muscle and the gills. The **nervous system** has been much modified by the sessile habit of life of the oyster. The cerebral ganglia are represented by a nerve ring, containing ganglia, which surrounds the mouth; it is called the **circumpallial nerve**. Fibers from this ring go to the pigmented sense papillæ at the margin of the mantle. The **visceral ganglia** lie along the antero-ventral side of the muscle and are joined with the cerebral ring by **longitudinal connectives**. The pedal ganglia are wanting.

PELECYPODA

A HARD-SHELL CLAM (*Venus mercenaria*)

This is a very common marine mollusk which inhabits the
sandy bottoms of the ocean along our shores. The soft-shell
clam (*Mya arenaria*), which lives in mud flats between tides,
resembles it very much in structure and may be used for this
dissection.

Study first the live animal, if possible. Its body is unsegmented and is entirely enclosed in a bilateral, bivalve shell,
which is the cuticula of the animal richly charged with calcareous salts. The two valves of the shell cover the right and
left sides of the animal and are joined together on its dorsal
side by the dark-colored **hinge ligament**, while their ventral edges
are open; the animal is thus very much compressed laterally.
The anterior end of the animal is truncated; the posterior end
is elongated. Which is the right-hand valve? The elevation
on each valve near the hinge ligament is called the **umbo**. It is
the oldest portion of the shell; from it as a beginning point
the shell has grown in size to its present proportions by addition to its ventral edge. Note the parallel lines of growth.
The ventral edges of the shell are thus the youngest portions
of them.

Exercise 1. Make a drawing of the right-hand valve, indicating
the anterior, posterior, dorsal, and ventral aspects, and
showing the lines of growth.

Exercise 2. Make a drawing of the dorsal aspect of the animal.

Kill the animal by immersing it for a few minutes in hot
water (70° C.). As the shell is kept closed by the contraction of

the two muscles which pass between the valves, it will gape open as soon as the animal is dead and the muscles are relaxed. It is the elasticity of the hinge ligament which causes it to open.[1]

Examine the animal as it lies in the shell. It will be seen that the inner surface of each valve is covered with a soft, slimy membrane, whose lower edge is parallel with the edge of the shell. This is the **mantle**; it is a double fold of the dorsal integument of the body, one side of which is covered by either fold. The mantle is the matrix of the shell, *i.e.*, it secretes it. The lower edge of the mantle is provided with muscle fibers and can be extended beyond the edge of the shell; it also possesses sensory functions; in some pelecypods eyes are situated in the mantle's edge.

Observe the large, soft **visceral mass** hanging between the two lobes of the mantle; it contains most of the viscera of the animal. On the lower side of the visceral mass, *i.e.*, towards the gape of the shell, is the muscular wedge-shaped **foot**, which can be extended beneath the edge of the shell and is the organ of locomotion. Do you see the two leaf-like **gills** on each side of the visceral mass and foot? Observe the two large **adductor muscles**, one in front of and the other behind the visceral mass, which pass from one valve to the other and serve to close them.

Pass a knife between the mantle and the left shell and separate them from each other. Cut the two muscles close to the shell; cut the hinge ligament and remove the left shell.

Study the inner surface of the shell. Note the two large scars marking the surfaces of attachment of the adductor muscles; just above the anterior scar is that of a much smaller

---

[1] The shell may also be opened by inserting some sharp, wedge-shaped instrument between the valves. The valves are thus pressed apart far enough to admit the blade of a scalpel, by means of which the adductor muscles should be cut close to the left valve of the shell. The hinge ligament should then be cut and the left valve be removed.

muscle, the anterior **retractor** of the foot. Note the broad line which joins the scars and runs parallel with the edge of the shell except near the posterior muscle scar, where it bends forward, forming a triangular indentation. This is the **pallial line**; it is formed by the insertion in the shell of the delicate muscle fibers near the edge of the mantle. The indentation is the **pallial sinus.** Note the hinge teeth just beneath the umbo.

Exercise 3. Draw a view of the inner surface of the shell.

Break the shell and examine the broken edge with a hand lens. Study the structure of the shell. It is composed of three layers — the inner **mother-of-pearl layer,** which is secreted by the entire surface of the mantle, the **prismatic layer,** and the **organic layer** or **periostracum** on the outside. The two latter layers are secreted by the edge of the mantle; the periostracum is very thin and gives the color to the shell. Place a piece of the shell in a solution of hydrochloric acid; note the effervescence which results; note also that an inorganic remnant, even of the two inner layers, is left.

Exercise 4. Draw a view of the broken edge of the shell on a scale of 5. Show the prisms of the prismatic layer.

Place the animal in water and study it as it lies in the right shell.[1] The two halves of the mantle will be seen to envelop entirely the visceral mass of the foot. Over the dorsal portion of the visceral mass the mantle is fused with it and cannot be separated, but the lateral and the ventral portions of the mantle lobes hang free, enclosing an extensive space, which is called the **mantle cavity.** In this cavity, on each side of the visceral mass, lie the two leaf-like **gills.** Observe the edges of the mantle. They are fused forward of the anterior adductor

[1] For the study of the soft parts of the clam it is well to have also at hand a specimen which has been deprived of both valves of the shell.

muscle; the entire ventral edges are free and permit the foot to protrude between them; their posterior edges are richly pigmented, and are also fused and modified to form the two **siphons**. These are protrusile tubes, through which water is taken into and expelled from the mantle cavity. Probe them. Note on each side below the posterior adductor muscle the triangular muscle which connects the siphons with the shell. It is the **siphonal retractor muscle**. Between the two siphons in the mantle cavity note the short transverse **septum** which divides the posterior portion of the mantle cavity into two chambers, a **dorsal** and a **ventral** one. The latter is the very large **branchial chamber**, which contains the visceral mass and the gills, the former, the very small **cloacal chamber**. The ventral siphon is called the **branchial** or **incurrent siphon**; through it the water streams into the branchial chamber bearing food and air for respiration. The dorsal siphon is called the **excurrent** or **cloacal siphon** and through it water passes outward from the cloacal chamber charged with carbon dioxide of respiration and with fæcal matter from the alimentary tract. Probe the cloacal chamber.

Carefully remove the left mantle lobe after cutting it with fine scissors at its line of attachment, beginning at the forward end. Cut off the siphonal muscle, leaving the siphon in position. Place the animal in water and study the arrangement of the organs. Observe the position of the **gills**; note in front of them two triangular flaps, the **oral palps**; in the median line between the two pairs of oral palps is the **mouth**; find it. Along the base of the gills note an elongated passage leading posteriorly to the cloacal chamber, the **suprabranchial passage** of the outer gill. Blow into this passage at its hinder end in the cloacal chamber with a blow-pipe, or probe it.

Observe again the siphonal region. Note the short septum which separates the branchial from the cloacal chamber, and the opening between it and the visceral mass: probe this opening. Just beneath the umbo will be seen through the semi-transparent body-wall a dark-colored mass, the **liver**, back of which are the yellowish **reproductive gland** and the dark-colored **organ of Keber.** Back of the latter is the **pericardium,** within which is the **heart.** Beneath the heart and in front of the posterior adductor muscle is the dark-colored **kidney.** Passing through the pericardium and the heart and above the posterior adductor muscle to the cloaca will be seen the **rectum.** It ends with the **anus** near the hinder surface of the muscle. Open the cloacal chamber by a slit in the side of its siphon and find the anus.

Exercise 5. Draw a semidiagrammatic view of the animal lying in the right-hand valve of the shell, representing the organs above mentioned. Carefully label all.

The respiratory system. The gills have already been noticed. The two gills on each side are, by way of origin, but a single organ, which is called the **ctenidium.** The clam is thus provided with a single pair of ctenidia, which are homologous to those of the squid and of snails. Each gill consists of a pair of plates or lamellæ united at their lower edges and open above, and further joined by vertical or dorso-ventral cross-partitions, the **interlamellar partitions.** The space between the lamellæ is thus divided into parallel, vertical chambers, the **water-tubes,** which run from the bottom to the top of the gill and open above into the **suprabranchial passage.** This is a wide canal running along the base of each gill to the cloacal chamber. The course of the suprabranchial passage of the outer gill has already been noted. In order to observe that of the inner gill, lift up both gills; the inner suprabranchial passage will be seen at the base of the inner gill. Probe from the cloacal chamber into it. Notice that back of the visceral mass the two inner suprabranchial passages coalesce and form a single passage.

Study the finer structure of the gills. Place a gill on a glass slide in a little water and with forceps and knife carefully separate the lamellæ. Mount a piece of a lamella in water and study it under a compound microscope. Note the vertical interlamellar partitions. Observe that the lamella is a delicate lattice work made up of ridges, the **gill-filaments**, which run vertically and thus parallel with the interlamellar partitions, and of cross-ridges, the **interfilamentary connections**, which run between and connect the vertical filaments. The apertures in the lattice work place the water-tubes in communication with the water of the branchial chamber. The gill-filaments are provided with cilia, as may easily be seen if the gill be alive, the action of which causes streams of water to pass into the water-tubes. The course of the respiratory water is from the branchial chamber into the water-tubes, through which it passes to the suprabranchial passages, and through these into the cloacal chamber, whence it is ejected through the cloacal siphon.

Exercise 6. Draw a diagram of the respiratory system showing the gills and their relation to the suprabranchial passages. Show the direction of the flow of the respiratory water by means of arrows.

Exercise 7. Draw a diagram showing the structure of a lamella.

The circulatory system. With fine scissors carefully cut open the pericardium by a slit along its dorsal border and expose the heart. Note the heart with the rectum passing through it. The heart consists of three chambers — a median, thick-walled **ventricle** and two lateral **auricles**. These latter are delicate, thin-walled organs, triangular in shape, the base of the triangle lying along the dorsal border of the gills and the apex communicating with the ventricle. If the left auricle has been injured in the dissection, the right one is easily seen by looking across the pericardial space. From the ventricle an anterior and a posterior artery pass to either end of the body. The posterior artery expands, near the posterior end of the pericardium, to form a large thick-walled sac, the **arterial bulb**. These two arteries lie alongside the rectum, to which the anterior one is dorsal and the posterior one is ventral; they are difficult to distinguish from it, except in specimens in which the heart has been injected.

The course of the blood is the following: by the contraction of the heart the blood is sent to all parts of the body, whence it is conveyed through lacunæ to the kidneys and thence to the gills; here it circulates in vessels which run through the inter-lamellar partitions, the gill-filaments, and the interfilamentary connections, and is purified; it then passes into the auricles.

The **excretory system** consists of a pair of **kidneys** which lie just beneath the pericardium and in front of the posterior adductor muscle. Each kidney consists of two parts, the **kidney proper** and the **ureter**. The former is a dark, thick-walled gland which lies beneath the ureter and communicates with it at its hinder end. The ureter is a thin-walled vessel lying above the kidney proper, with a small external opening in the side of the visceral mass near the base of the inner gill. Cut off the gills and look for the external opening; it may be recognized by its white lips. The kidney also possesses at its anterior end a duct leading into the pericardial cavity. Slit open the ureter and kidney proper and observe their inner structure.

Exercise 8. Draw a diagram representing the pericardial cavity and the kidney, showing the relation of the two structures to each other. Draw the heart in the pericardial cavity, showing the relation of the auricle to the gills.

The digestive system. Find the **mouth** between its two pairs of palps and place a bristle in it; note the **upper** and the **lower lips**, which connect the upper and the lower pair of palps, respectively. The mouth is seen to the greatest advantage in a specimen which has been taken out of both shells. Trace the rectum from the anus through the heart to the point where it meets the visceral mass. With forceps and knife carefully remove the tough white integument which covers the left side of the visceral mass. The soft cream-colored mass filling the greater part of it is the reproductive gland, the greenish mass above is the liver. Imbedded in these masses lies the **alimentary tract**, a narrow, delicate tube, which will be injured in the dissection unless the greatest care be taken. Beginning with the mouth gently scrape away the soft mass which surrounds the alimentary tract, laying it entirely bare. The water in the dissecting pan must be frequently renewed to keep it clear, and great care taken not to break the canal. The mouth opens into the short **œsophagus**, after which the canal dilates to form the **stomach**. The liver surrounds the stomach and is connected with it by several ducts. Back of the stomach is the **intestine**, which first runs backward and downward to the posterior part of the visceral mass, after several turnings in the lower part of which it bends upward and runs forward parallel with the posterior margin of the visceral mass to its dorsal border, where it leaves it. Here the **rectum** begins and passes through the heart and above the posterior adductor muscle to the **anus**. A small transparent rod is often present in the intestine; its function is unknown.

Clams feed upon minute organisms and organic particles contained in the water. Some of the water in the mantle cavity is drawn into the mouth by the ciliated oral palps and passes through the alimentary tract, where the organic substances are digested and absorbed.

Exercise 9. Draw a diagrammatic view of the digestive system.

The reproductive system. The sexes are separate. The **reproductive glands** (testis or ovary) are very similar to each other and consist of a pair of cream-colored masses which fill a greater part of the visceral mass. Their external openings are a pair of minute pores, one on each side of the visceral mass just below and in front of the opening of the ureter. They can often be located by pressing out from them eggs or sperm.

The **nervous system** consists of three pairs of ganglia — the **cerebral ganglia** or **brain**, the **pedal ganglia**, and the **visceral ganglia**, and the nerves proceeding from them ; each of the last two pairs is joined with the brain by a pair of nerve-connectives.

First find the visceral ganglia. They are a small pinkish mass on the ventral surface of the posterior adductor muscle, with nerves radiating in all directions. Two of these nerves, the **cerebro-visceral connectives**, will be seen passing forward, one on each side of the visceral mass.

Find next the brain. It consists of a pair of pinkish ganglia situated above the mouth, just behind the anterior adductor muscle. The two ganglia are not so close together as are those of the visceral pair ; they lie on each side of the muscle and are united by a commissure. Each ganglion sends out three large nerves — the **cerebro-visceral connective**, which goes to the visceral ganglia, the **cerebro-pedal connective**, which goes to the pedal ganglia, and the **pallial nerve**, which passes to the mantle. Find these nerves.

The pedal ganglia form a nervous mass buried in the foot near its base. They must be sought by cutting into the foot near its base and may be recognized by their pink color. In contact with them is a sense-organ called the **otocyst**.

Exercise 10. Draw a diagram representing the nervous system.

Make several transverse sections with a razor through the region of the heart of a clam which has been previously hardened. Identify all the organs which appear.

Exercise 11. Draw a diagram representing a cross section ; carefully label all the organs.

GASTROPODA

A PULMONATE GASTROPOD. A LAND SNAIL (*Helix pomatia*)

This snail is very common in Europe, in many parts of which
it is used for food. It is imported into this country for the
same purpose and may be obtained at small cost in New York
and Philadelphia. It is especially adapted for dissection, but
any large Helix may be used instead. The large slug (*Limax
maxima*) is very similar to Helix in structure and may also be
used, but as it has no coiled shell that feature of the dissection
would be omitted.

The snail is a terrestrial animal and feeds principally upon
leaves. It hibernates in the winter under stones and logs after
having first closed the mouth of its shell with a thin disc of
calcified slime called the **epiphragma**. If it is still in winter
quarters, when obtained, the epiphragma should be removed
and the animal placed among fresh leaves in a warm room,
when it will soon come out of its shell and begin to feed.
Snails are best killed for dissection by drowning. They should
be placed in a large covered jar of water, when they will die
expanded in from one to two days. If the air be first boiled
out of the water the process will be accelerated, but the animal
should not be placed in water which is still hot.

Study the external characters of the animal. Its body is
unsegmented and is covered with a shell, but unlike the shell
of the pelecypod that of the snail is a univalve. As in
other mollusks, the shell is the cuticula of the animal charged
with calcareous salts, and forms an exoskeleton. In shape the
shell is an elongated cone which has been twisted to the right,
forming a closely coiled spiral. The tip of the spiral is called

the **apex**, the opening is called the **mouth**, and its axis, the
columella. How many turns does the spiral make? The apex
corresponds to the umbo of the lamellibranch; it is the oldest
part of the shell, the point from which its growth has proceeded.
Note the parallel lines of growth. The ventral edge or mouth
of the shell is thus its youngest part. The animal can with-
draw its entire body within the shell, but when it is walking
or feeding it protrudes its **head** and **foot**. The **visceral mass**, however, containing all of its viscera, is always covered by the shell
and has thus its exact shape, *i.e.*, it is an elongated cone which
has suffered a dextral twisting so as to form a closely coiled
spiral. As a matter of fact, however, it is the visceral mass
which has been primarily twisted; the shell is twisted because
it covers the visceral mass. If the spiral were to be imagined
uncoiled and extending straight up above the foot, the apex
would be the uppermost and the foot the lowermost portion of
the body; the apex is thus, morphologically, the dorsal and the
foot is the ventral aspect of the animal.

As in the pelecypod, the visceral mass is enclosed in a
mantle, which is a fold of the dorsal integument, but unlike the
pelecypod it is a single fold and not a double one. This
fold falls about and covers the visceral mass on all sides, as
does a thimble the finger it is on, and secretes the shell on its
outer surface. The ventral edge of the mantle is provided
with muscles, so that it can be protruded beyond the mouth of
the shell or retracted within it. This edge is called the **collar**.
Find it in your specimen. On the right side of the animal note
the deep notch and the round hole in the collar. This is the
respiratory pore, which opens into the **respiratory chamber**. This
chamber is the **mantle cavity**. Probe it gently and determine
its extent. The animal being terrestrial has no gills, but
respires by means of a **lung**, which is a highly vascularized
portion of the wall of the mantle cavity. In a live animal note
its power to open and close the respiratory opening.

The **foot** of the animal forms a broad creeping disc, adapted for locomotion on flat surfaces. Its wave-like undulations may be observed by causing the animal to walk over a glass plate. The **head**, which is wanting in the pelecypods, forms the anterior end of the animal and bears two pairs of hollow, retractile tentacles, the posterior pair carrying each an **eye** at its extremity. The **mouth** of the animal is between and a little below the base of the anterior pair of tentacles. Probe it and note the paired lobed **lips**. Just beneath the mouth is the broad opening of the **pedal slime gland**. Probe it and note the extent of the gland. On the right side of the head is a straight groove which extends to a depression just behind the base of the anterior tentacle. This depression is the **common genital pore**, the animal being hermaphroditic. The **anus** is a small opening just beneath the respiratory pore at the end of a deep groove. It is not easily observed from the outside.

Note the asymmetry of the animal. Its spiral twist has been the cause of the loss of the primitive bilateral symmetry of the visceral mass and shell. They are not borne squarely above the foot, but obliquely and to the left. The respiratory pore (*i.e.*, the opening of the mantle cavity) and the anus have not a median posterior position, as must have been the case in the primitive ancestor of the animal, but have suffered displacement to the right side. Other instances of asymmetry will be noticed as the dissection proceeds.

Exercise 1. Draw a side view of the animal seen from the right side as it appears when it is moving and when the head and foot are out of the shell, and label the parts above mentioned.

Exercise 2. Draw a similar sketch of a front view of the animal.

Remove the dead animal from its shell in the following way: place it for five minutes in strong alcohol, or for half a minute in very hot (not boiling) water, in order to loosen the shell; twist it then out of the shell; this must be done very gently, otherwise the animal will be torn.

Exercise 3. Draw the shell showing its opening on the right.

Break off a portion of the edge of the shell and examine the broken edge with the aid of a hand lens. Note the three layers which compose the shell—the inner **pearly layer**, which has been secreted by the entire surface of the mantle; the thick **middle layer** and the thin **outer layer** or **periostracum**, which have been secreted by the collar. The periostracum is a horny, uncalcified layer, which gives the color to the shell.

The internal organs. Take the snail, deprived of its shell, in the hand, and, remembering that the outer side of each whorl of the spiral is on the left side of the animal and that the inner side of the whorl is on the right, observe the extent of the mantle cavity. Put the blow-pipe through the respiratory pore and blow into the mantle cavity. It will be seen to extend from the collar to the posterior side of the first whorl. Examine the mantle wall with a hand lens and against the light. The network of blood vessels will be seen, which constitutes the **lung**. On the hinder border of the mantle cavity note the **kidney**, an elongated, light-colored, triangular organ; just in front of it and beneath it, *i.e.*, between it and the mantle cavity, is the **heart** within the **pericardium**; note the two chambers of the heart, the dorsal **auricle** and the more ventrally placed and larger **ventricle**. Back of the kidney is the dark-colored **liver**, which, with the **intestine** and the light-colored **reproductive tract**, occupies the remainder of the coils of the spiral. Note the **rectum**, a broad tube on the inner (right) border of the mantle cavity going to the **anus**. Cut a small hole in it, and through this pass a probe to the anus.

The mantle cavity. Lay this open in the following way: with fine scissors cut through the collar at the respiratory pore;

then make an incision in the mantle wall from this opening
following the collar round the outer side of the whorl to the
heart; continue the incision across the artery leading out of
the heart, and through the delicate membrane between the liver
and the kidney to the rectum, at the inner border of the whorl.
The mantle can now be laid back and its cavity with the organs
exposed. The broad rectum will be seen running along the
entire inner border of the mantle cavity. Make, now, an additional incision from the respiratory pore along the inner (lower)
border of the rectum as far as the kidney. Lay back the mantle
and pin it down as flat as possible under water. Identify the
heart within the pericardium, the kidney, and the rectum.

The respiratory and circulatory systems. Observe the **lung**, the
network of blood vessels in the inner surface of the mantle, and
the large **pulmonary vein**, which runs along the kidney to the
heart. Slit open the pericardium. The two chambers of the
heart will be more distinctly seen, the thin-walled **auricle** into
which the vein runs and the larger **ventricle**. Back of the latter
the **aorta** passes into the viscera; its cut end will be seen.

The process of respiration and circulation is the following:
the air is drawn into the mantle cavity through the respiratory
pore; this is accomplished by the alternate enlarging and con-
tracting of the cavity by means of the muscular body-wall which
constitutes its floor. Notice the longitudinal and the transverse muscles in this floor. The blood circulating in the lung
is oxygenated and passes into the heart through the pulmonary
vein as arterial blood. It is forced by the heart through the
aorta, and thence through arteries to all parts of the body,
whence it returns through blood lacunæ to the lung.

The excretory system. The large **kidney** has already been seen.
It is a sac, the glandular projections of the walls of which
almost fill its lumen. As is the case with pelecypods, the
kidney communicates with the pericardial space through a fine
canal and also with the mantle cavity by means of a **ureter**.

The pericardial canal is opposite the ventricle and cannot be seen easily. The ureter may be easily traced. It is a wide canal which leaves the kidney at its forward end near the place where the pulmonary vein approaches the kidney; it first runs along the inner side of the kidney to its hinder end; here it doubles on itself and passes forward to the inner edge of the mantle, where it runs beside the rectum to a point near the respiratory pore and opens into the mantle cavity.

It will be noticed that the heart and the kidneys are both asymmetrical organs. The heart has but one auricle; it will be remembered that in the pelecypod the auricles are paired organs; one of the pair must thus be wanting in the snail. There is also only one kidney and one ureter, instead of a pair of each, as in the pelecypod. It is the left member of the pair in each case which is wanting.

Exercise 4. Draw a view of the inner surface of the mantle on a scale of 3, showing the organs mentioned above; label all.

The digestive system. Pass a bristle through the anus into the rectum in order to mark it. With two strong pins firmly fasten the extreme forward end of the animal's foot and also its hinder end to the wax of the dissecting pan. With sharp, fine scissors cut through the floor of the mantle cavity and the collar in the median line; carry the incision forward in the median line along the head between the base of the tentacles to the mouth. Care should be taken in making this incision not to cut the organs beneath. Spread the flaps as widely as possible to the right and left and pin them down, exposing thus the organs in the forward part of the body.

The white organs on the right side of the body belong to the reproductive system. The large dark organ in the center, or on the animal's left, is the **stomach**. Find the slender curved **œsophagus** which leads forward from it to the dorsal side of the large muscular **pharynx**. The œsophagus is encircled by the white nerve collar, the dorsal portion of which is the brain. If, however, the animal died in a retracted condition the pharynx may have slipped back through the nerve collar, which would then encircle the forward end of that organ. Note the two white, leaf-like **salivary glands** which lie close against the wall of the stomach, and trace their **ducts** forward to the pharynx. Lying above and across the œsophagus is the white cylindrical penis, which will be seen to extend from the genital pore at the right of the mouth and to bend sharply on itself. The bend of the penis is connected by a long retractor muscle to the dorsal body-wall. Find it; cut it and pin the penis on the animal's right. Note the broad, glistening retractor muscle connecting the pharynx with the ventral, posterior body-wall. Notice its shape; with strong forceps pull it loose from the pharynx and entirely remove it. Also beneath it, note the still larger retractor muscle running from the forward end of the head back to the same locality. What is the function of these different retractors? The dark-colored sheaths and the retractor muscles of the tentacles will also be seen; trace these muscles to their origin. Find the nerve which passes from the brain into each tentacle.

Separating the delicate filaments connecting the stomach with the surrounding organs, and pushing it and the œsophagus to the animal's left, find the large nervous mass which forms the ventral portion of the nerve collar, and the nerves radiating from it. It is an agglomeration of ganglia, being made up principally of the **pedal** and the **visceral ganglia**. Press the reproductive organs to the animal's right and pin them down. The **receptaculum seminis**, a small spherical body the size of a shot, at the end of a long tube, will be found in a bend of the intestine, from which it must be separated.

We turn now to the other end of the **intestine**. Trace the **rectum** from the **anus** to the point where it is surrounded by the **liver** and carefully dissect away the integument which covers the inner surface of the whorl. The light-colored **hermaphroditic gland** will be exposed. Remove, then, the delicate integument which covers the outer surface of the whorl, and the dark-brown **liver** will be exposed. Press the liver away from the intestine and completely free it, without, however, breaking either liver or intestine. Great care should also be taken not to injure the hermaphroditic gland, which is the yellowish mass on the inner side of the last whorl, or the **hermaphroditic duct** leading away from it. Note that the liver is composed of two masses, the smaller of which is of spiral form and occupies the apex of the shell; the larger is subdivided into three lobes. Note also the two main **bile ducts** which join the liver with the intestine. The **visceral artery** will be seen lying upon the liver, sending branches off on both sides, and must not be confused with the bile ducts, which it resembles in appearance. It carries blood from the aorta to the top of the spiral, supplying all the organs of the visceral mass. At the point where the bile ducts communicate with the intestine that organ makes a sharp turn.

Spread out the digestive tract to the animal's left and pin it down, without, however, removing or breaking the hermaphroditic gland or duct. The stomach will be seen to extend nearly to the liver. It is succeeded by the intestine, which soon makes the sharp turn above mentioned, receives the bile ducts, and passes into the rectum at the right side of the mantle cavity.

Exercise 5. Draw an outline of the alimentary tract from the mouth to the anus on a scale of 2 and label all its parts.

Study the structure of the pharynx. Pass a probe into the mouth and notice the extent of the pharyngeal cavity. Notice the transverse horny **jaw** in the roof of the mouth. With a sharp knife split the dorsal pharyngeal wall, taking care not to injure the nerve collar or reproductive organs. Notice that the connection of the œsophagus and the salivary ducts with the pharynx is near the dorsal wall of the latter organ. Observe the thick muscular **tongue**, the organ by means of which the animal grinds its food. Its surface is covered with a ribbon set with small teeth, called the **radula**. This can be easily pulled off with forceps. Mount it on a slide in water or glycerine and study its surface under a high power of the microscope.

Exercise 6. Make a drawing of several of the teeth.

The reproductive system. The snail is hermaphroditic, but is not self-fertilizing. The **hermaphroditic gland**, which at different times produces both spermatozoa and ova, is situated on the inner side of the smaller lobe of the liver. The **hermaphroditic duct** is a delicate, white, convoluted tube, which goes from the hermaphroditic gland to the **albuminous gland**, a large white body lying near the liver. From this organ the **oviduct** and **vas deferens** pass forward to the genital opening near the mouth. These canals are side by side and connected with each other for the first part of their course, but may be distinguished by the character of their walls, the oviduct having folded glandular walls, while the vas deferens is a narrow tube with thin walls. It is through the latter canal that spermatozoa pass out from the hermaphroditic duct, while the ova pass out through the oviduct, the glandular walls of which, together with the albuminous gland, secrete the albumen which surrounds them when they are extruded. Near their forward end these two canals separate. The oviduct loses its glandular walls, becomes cylindrical in shape, and expands to form the **vagina**. This is a thick-walled vessel with which are connected the following accessory genital organs: the **receptaculum seminis**, a small spherical organ, already mentioned, lying in the bend of the intestine and joined with the vagina by means of a long tube which lies along the oviduct; the **mucous glands**, two bunches of tubular glands; and the **dart-sac**, a thick-walled sac which contains a **calcareous spicule**. Identify these organs.

The **vas deferens**, after separating from the oviduct, passes under the retractor muscles of the tentacle to the distal end of the **penis**. This organ has already been noted; it is tubular in shape and lies in a bent position across the œsophagus. A retractor muscle inserted at the bend connects it with the dorsal body-wall. At the point where the vas deferens meets it is the **flagellum**, a long, tubular sac into which spermatozoa pass from the vas deferens and where they are massed together to form **spermatophores**. Both penis and vagina communicate, side by side, with the **genital cloaca**, which opens to the exterior through the common genital pore.

When two animals pair each receives a spermatophore from the other. This passes into the receptaculum seminis, which is thus filled with the spermatozoa of the other animal, and these finally fertilize the eggs as they pass into the vagina from the oviduct.

Exercise 7. Make a semidiagrammatic drawing of the reproductive organs on a scale of 2.

Split the dart-sac and take out the dart; mount it on a slide in water or glycerine and examine it under a compound microscope.

Exercise 8. Draw the dart.

The nervous system. Sever the œsophagus and remove the reproductive and digestive systems, leaving the pharynx in the body and taking care not to injure any of the nerves. The principal ganglia are contained in the **nerve collar**. The two **supraœsophageal ganglia**, which constitute the brain, will be seen joined by a broad **transverse commissure**. From their anterior surface nerves run to the tentacles, and from their inner surface a pair of nerves runs to the posterior end of the pharynx, where they meet a pair of small **pharyngeal ganglia**. The **supracœsophageal ganglia** are connected by broad connectives with the **subcœsophageal ganglia**. Remove the pharynx from the body. By slightly scraping the subcœsophageal ganglia with a small scalpel, it will be seen to consist of two principal ganglionic masses. The forward mass is a pair of ganglia, the **pedal ganglia**; the hinder mass consists of the large **visceral ganglia**, at the side of which is the pair of small **pleural ganglia**. Observe the nerves radiating from the subcœsophageal ganglia, and determine so far as possible to what organs they go.

Exercise 9. Make a semidiagrammatic drawing of the nervous system.

Organs of special sense. The **eyes** of the snail at the end of the posterior tentacles have already been noted. They are easily seen in a large animal which has its tentacles extended. The snail is also provided with a pair of **auditory organs**. They consist of two small sacs imbedded in the pedal ganglia. In order to see them cut off the subcœsophageal ganglion, mount it in glycerine and examine it under a compound microscope. The auditory nerves are very delicate and come from the supracœsophageal ganglia.

CEPHALOPODA

A DIBRANCHIATE CEPHALOPOD. A SQUID (*Loligo pealii*)

The squid is a very common marine animal. It is social in its habits and swims about in large schools in search of its food, which consists of crustaceans, small fishes, etc. When alarmed by the presence of its natural enemies, which are many kinds of fishes, it clouds and darkens the water by ejecting into it an ink-like fluid. The fresh animals are studied with greater profit than those which have been preserved in alcohol, as this changes the nature and appearance of many of the organs; if they must be preserved, formalin should be used.

External anatomy. Observe the cylindrical, bilaterally symmetrical body; at one end is a pair of broad **fins**, and at the other, the movable **head** bearing ten **arms**, two of which are much longer than the others. The **mouth** is at the base of and surrounded by the arms, and the brown horny **beak** may usually be seen protruding partly from it. The large **eyes** are on the sides of the head at the base of the arms. Each is covered by a cornea, which is pierced by a small hole between the eye and the base of the arms, so that sea water is admitted freely into the space between the cornea and the pupil, and may take the place of the aqueous humor of the vertebrate eye. A transverse fold on the side of the head between the eye and the body is the **olfactory organ.** Observe the pigment spots or **chromatophores** which are distributed over the body; they are constantly changing in shape and size during life, causing corresponding changes in the color and appearance of the animal.

The head and neck project from the large **mantle cavity**, into which they can be partially withdrawn by means of powerful retractor muscles, in very much the same way that a snail's head and foot can be withdrawn into its shell. The **siphon** or **funnel**, a large funnel-shaped organ at the base of the head, also projects from it and can be similarly withdrawn. Gently probe the mantle cavity and determine its extent. The **mantle** constitutes the outer surface of the body. It will be seen to be a cylindrical structure with thick, muscular walls, within which lie all the viscera of the animal; its free edge is called the **collar**, as in the snail. It is also necessary to observe that the mantle is not a paired organ, as it is in the clam, but an unpaired one as in the snail. The squid has no foot, as has the clam or the snail, but morphological equivalents of the foot are present in the arms and the siphon.

Since in all mollusks the foot or its equivalent occupies a ventral position, and the visceral mass a dorsal position, the arms of the squid, together with the head, must be on its ventral side, and the opposite end with the broad fins must be dorsal; the animal is thus enormously extended dorso-ventrally. It will be readily seen also that the mantle falls as a cylindrical fold from the dorsal end about the entire body, exactly as it does in the case of the snail. In fact, if the coils of the snail's visceral mass could be straightened out, the mantle would fall as a cylindrical fold from its dorsal end and terminate in the collar below, in the same way as in the squid. The morphologically posterior side of the animal is that on which the siphon is situated, the anterior side is the opposite one. In common parlance, however, the head end of the squid is called the forward end, and the fin-bearing end, the hinder. The side bearing the fins is likewise called the upper side or back, and the opposite side, on which is the siphon, the under or lower side. These terms, although incorrect in a strictly morphological sense, are much more convenient for general use and will be employed hereafter in these directions.

The mantle of the squid does not secrete an external shell as does that of the snail and the clam; in a long pocket on the upper side, however, is an elongate, horny structure, called the **pen**, which is secreted by the mantle and is the equivalent of the shell of other mollusks.

Make a short shallow incision in the upper surface of the mantle, beginning with the collar. Turn the flaps aside and note the brown, horny **pen** lying beneath. Do not remove it at present, as the dissection of the parts beneath might be disturbed by its removal.

Exercise 1. Make a drawing of the underside of the animal.

Note that the arms may be divided into a right and a left group, each containing five arms. Observe a single arm; how many rows of suckers has it? Observe the structure of a sucker. Note the difference between the two long arms and the others in the place of origin and the arrangement of the suckers.

The mantle cavity. Open the mantle cavity by a longitudinal incision through the thick mantle wall of the under side of the body to one side of the median line, running from the collar to the apex of the animal, taking care not to injure the delicate organs within. Notice, in the first place, that the collar is not attached to the head at any point of its circumference; and also that on the inner surface of the mantle, on the upper side of the body in the median line and also on each lateral surface, there is an elongate, cartilaginous structure which fits into a corresponding cartilage on the body, an arrangement which enables the collar to be applied very closely to the head.

Place the animal in water with the head away from you and pin down the flaps of the mantle. Observe the soft **visceral**

mass within it, and notice that it is fused with the mantle only in the median line of the back; also that the pen, which is imbedded in the mantle, protects the viscera on that side. Observe the **siphon** and probe it. It will be seen to be a funnel-shaped tube communicating between the mantle cavity and the outside. Slit it open and observe the flap-like **valve** at the forward end. Notice the **lateral pockets** on each side of the siphon which open toward the mantle cavity and occupy the space between the siphon and the median line of the back. They are separated from the siphon by the lateral cartilaginous rods above mentioned. It will be seen that while water can easily pass into the mantle cavity from the outside all around the neck, a contraction of the muscular wall of the mantle would force the water out through the siphon only, as that which is forced into the lateral pockets would at once swell them out and close the spaces at the sides of the siphon. It is, in fact, by thus shooting the water in the mantle cavity forcibly through the siphon that the animal swims.

Note the two large **retractor muscles** of the siphon and beneath them the two larger **retractor muscles** of the head.

Observe again the **visceral mass**; it is covered by a thin, transparent membrane, **the body-wall,** the extreme thinness of which is correlated with the thickness of the mantle which covers it. If the animal be a female that fact may be known by the presence of two very large, transversely striated bodies, called the **nidamental glands,** which lie near the center of the body, and are a part of the reproductive system. Carefully remove these in order to expose the organs beneath. If the animal be a male (and the student should obtain a male if possible), it can be recognized by the absence of nidamental glands and also by the presence of the **testis,** a large white tubular organ which lies near the median line toward the hinder end of the animal. In the female the **ovary,** which occupies a similar position, is often very full of the granular ova.

Notice in the mantle cavity the pair of plumose **gills** to the right and the left of the visceral mass, each attached to the inner surface of the mantle by a **mesentery.** Between the retractor muscles of the siphon and extending from the base of the gills forward to the siphon is the **rectum,** which terminates in the **anus,** with its two projecting **valves.** Find the valves. Beneath the rectum is the **ink-bag,** and both are attached to the organs beneath them by a **mesentery.** The ink-bag communicates with the rectum by means of a duct which joins it near the anus; this duct may be found by slitting the rectum for a short distance back of the anus, when the small opening may be made to appear by squeezing the ink-bag and forcing the ink into the rectum. Together with the fæcal matter from the intestine and other waste products, the ink is voided into the sea water through the siphon; its function is to cloud the water and thus hide the animal from its enemies. In the male animal notice the long, tubular **penis** to the right of the rectum (the animal's left); if the animal is a female, the thick-walled **oviduct** will be seen in a corresponding position.

At the base of each gill note a round disc-like body; this is a **branchial heart,** from which blood is sent into the gills; near each branchial heart, toward the median line and running forward alongside the rectum is an elongate, transparent structure, the **kidney.** The position of the kidneys may be determined by the two conspicuous white veins — the **precaval veins** — which pass through them longitudinally from one end to the other. These veins are wide spongy-walled structures which run to the branchial hearts and will be seen toward the median line from those organs. Just beneath the base of the two kidneys and between the branchial hearts is the **median** or **systemic heart,** into which blood pours from the gills. Note a median artery, the **posterior aorta,** which leads back from the systemic heart; it branches into three large **mantle arteries,** two of which pass to the right and left, respectively, and enter the mantle at the side, while the other passes into the mantle in the median line; it is through these arteries that the mantle is supplied with blood.

On each side between the base of the gill and the rectum and extending parallel with the latter organ, notice again the delicate **kidney**; each of the pair of kidneys extends backward to a point a short distance back of the branchial heart, and forward to a point back of the base of the ink-bag, where it communicates with the mantle cavity through a small opening. Find the two openings by lifting up the body-wall with forceps and blowing on it with a blow-pipe, when they will appear.

Running back from the branchial heart on each side is a wide vessel, the **postcaval vein**; the forward end of this vein has thick, spongy walls like those of the precavals and is easily seen; the greater part of it, however, has extremely thin walls and can be seen with difficulty. Near the base of each gill note also a vessel which runs forward and laterally into the mantle; this is the **mantle vein**. Just back of this vein is a muscle which connects the gill with the mantle; it is the **branchial retractor muscle.**

Note the two large **stellate ganglia** in the forward part of the inner surface of the mantle, and the radiating nerves which each ganglion sends into the mantle.

In the hinder portion of the visceral mass in the male animal observe on the animal's left (the observer's right), just behind the branchial heart, a coiled tube, the **vas deferens,** and in the female the thick-walled **oviduct.** Extending farther back and near the median line is the large white **testis** in the male and the large **ovary** in the female.

Exercise 2. Make a large sketch of the mantle cavity of the animal showing these organs, and label all.

With fine scissors and forceps carefully dissect away the delicate transparent body-wall and expose the organs beneath, taking care not to injure them.

The excretory system. The **kidneys** and their external openings have already been observed. As in other mollusks, the kidneys also communicate with the pericardial space.

The circulatory and respiratory systems. Pushing aside the organs which partly conceal it, observe again the **systemic heart**; note its shape and slightly asymmetrical position. Extending from its forward end is the **anterior aorta**, which takes blood to the forward part of the body; its course cannot be followed at present. The hinder part of the body is supplied with blood by the **posterior aorta**. This vessel, as we have already seen, leaves the hinder end of the systemic heart; it sends off two pairs of small **arteries** to the stomach and to other viscera, and then branches into the three **mantle arteries** already mentioned. Find them all and trace them as far as possible. Observe again the two **branchial hearts**. Note the **branchial artery,** by which blood passes from the branchial heart to the gill; also the **branchial vein,** through which it passes into the systemic heart.

Observe again the veins which bring the blood to the branchial hearts. The **precavals** bring blood from the forward part of the body. Trace them forward. They enter the kidneys near the forward end of those organs and traverse their glandular walls back to the branchial heart. Press aside the rectum and the forward end of the kidneys, and observe where the two **precavals** come from beneath and enter the kidneys. With fine scissors cut the connective tissue which binds the veins, and also the mesentery which holds down the rectum and the ink-bag, and turn these organs back. Trace the two **precavals** forward; they will be seen to come from a delicate **median vein** which may be followed into the head. Observe again the **postcaval veins,** which bring blood from the hinder part of the body and join the branchial hearts near the same place as the precavals. Their forward ends also traverse the glandular walls of the kidneys and are here conspicuous; back of these they are much wider, but are very thin-walled and not easily seen. Trace them as far as possible. Observe again the **mantle veins**, which bring blood from the mantle to the branchial hearts.

The course of the blood is the following: it enters the branchial hearts through the postcaval, precaval, and mantle veins; the contraction of these hearts sends it into the branchial arteries which pass along the upper side of the gills; it then traverses the delicate transverse filaments of the gills and becomes oxygenated, when it collects again in the branchial veins on the opposite side of the gills; through these it passes to the systemic heart, whence it is sent through the anterior and posterior aortas to the different parts of the body.

Exercise 3. Make a diagrammatic drawing of the circulatory and the respiratory systems.

The digestive system. Remove the kidneys and precaval veins. Beneath them will be seen a large glandular bilobed organ of somewhat doubtful function, called the **pancreas**. At its forward end a pair of cylindrical organs, the **liver ducts**, will be seen entering it from the **liver**. The pancreas is made up of anastomosing glandular projections of the walls of these ducts. Remove the gills, branchial hearts, systemic heart, and hinder arteries. The delicate body-wall should be completely removed from the entire visceral mass, and great care be taken not to injure the **stomach pouch** beneath. This latter organ is a large bag with thin transparent walls which extends to the extreme hinder end of the body; beneath it will be seen the large **testis** or **ovary**, according to the sex of the animal. This pouch is not really a part of the stomach, notwithstanding its name, but is a reservoir for the secretions of the liver, which communicates with it through the liver ducts. Carefully loosen the stomach pouch without separating it from the body and let it float in the water of the dissecting pan. It communicates with the thick-walled **stomach**, which lies just in front of it, but food substances are prevented from passing into it from the stomach by valves. Loosen the stomach, noticing that it is bound to the ovary or testis by an artery, the **genital artery**. At the forward end of the stomach are the **intestine** and the œsophagus, side by side; the former passes between the two halves of the pancreas and ends with the **rectum**; the œsophagus goes forward side by side with the anterior aorta to the middle of the large liver and passes through it in company with the aorta. The œsophagus is easily found by turning the stomach over. A small **ganglion** with radiating nerves will be seen by the side of the œsophagus near its junction with the stomach.

The **liver** is an elongated body lying between the retractor muscles of the head and of the siphon; two ducts emerge from it and pass through the pancreas to the stomach pouch. Loosen and remove the connective tissue around the liver and raise it up; the œsophagus and the aorta will be seen to pass through it towards the back of the animal and then forward to the head.

Remove the siphon and split the wall of the head; trace the **œsophagus** to its forward end. It will be seen to pass through the ganglionic mass which constitutes the central nervous system, and which is surrounded by a hard cartilaginous capsule. Forward of this it meets and ends in the bulbular **pharynx**. Near the forward end of the liver, and resting upon the œsophagus, will be seen the **median salivary gland**, the duct of which may be traced to the pharynx; near the hinder end of the pharynx is a pair of smaller **salivary glands**, which also communicate with it. Trace their ducts to the end. The alimentary canal will thus be seen to consist of the following organs: the muscular **pharynx**, with which a pair of small **salivary glands** and a single large **salivary gland** communicate; the narrow œsophagus; the thick-walled **stomach**; the **stomach pouch**, which communicates with the stomach by a valved opening; the elongated **liver**, which communicates with the **stomach pouch** by two long **ducts**; the bilobed pancreas; the intestine, which leaves the stomach near the point where the œsophagus enters it; the rectum, which is joined by the ink-bag and passes to the anus.

Exercise 4. Take the alimentary tract out of the body, pin it down, and make a drawing of it; label all its divisions.

Slit open the stomach and examine its ridges. Slit open the pharynx on the upper side; note the large chitinous jaws and the radula. The latter organ, like the radula of snails, is used in chewing the food; examine its surface under a microscope and note the calcareous teeth.

Exercise 5. Make a drawing of the jaws.

Exercise 6. Draw several of the teeth of the radula.

The reproductive system; the male. The principal genital organs have already been observed. The single median testis is a large, flat organ, dorsal to the stomach pouch, in the hinder portion of the visceral mass; the genital artery joins it with the surface of the stomach. The testis has no direct connection with the vas deferens, but is surrounded by a thin transparent membrane within which it lies as in a capsule, and into which the spermatozoa escape. The vas deferens, which is also unpaired, communicates with this capsule. It is a long and much-twisted tube with several wide glandular regions, and lies, bound by connective tissue into a compact mass, on the left side of the viscera. Take the entire system out of the body, put it in water, loosen and straighten out, so far as possible, the convolutions of the vas deferens. Beginning with its hinder end we find first a narrow, convoluted tube, then follows a thicker tubular portion, the vesicula seminalis; near the forward end of this portion is a glandular body, the prostate gland, and a membranous sac; a long, straight, narrow portion comes next, which widens to form the spermatophoric sac, within which the spermatophores are formed; then follows the tubular **penis**, which forms the forward end of the tract and has already been observed lying in the mantle cavity to the left of the rectum.

Exercise 7. (*a*) Make a drawing of the male genital tract.

Exercise 7. (*b*) Open the spermatophoric sac and look for spermatophores ; they are slender, white objects about half an inch long. Mount several on a slide and make a drawing of one.

The female. The single **ovary**, like the testis, is a large elongated gland occupying the hinder end of the visceral mass and surrounded by a **capsule**. The **oviduct** communicates with this capsule; it passes forward along the left side of the visceral mass, its walls becoming thickened in its course to form the **oviducal gland,** and opens into the mantle cavity by means of a large thick-lipped aperture to the left of the rectum.

Two pairs of prominent accessory glands are present in the female, the large, white, finely striated **nidamental glands,** which cover up most of the other organs of the visceral mass, and beneath them the much smaller **accessory nidamental glands,** which are pink-colored in life and lie to the right and left of the rectum; both pairs of glands open at their forward ends into the mantle cavity. These glands secrete the egg-capsules which protect the eggs after they are laid, and while development is going on within them.

Exercise 7. (*c*) Make a drawing of the female organs.

The nervous system. In the position of the principal ganglia the squid resembles the snail, but these ganglia are difficult to observe in a dissection because they are compactly massed together and are protected by a cartilaginous capsule which forms a sort of skull. The cerebral or **supraœsophageal ganglia** form a large mass above the œsophagus; broad commissures join it with the subœsophageal mass, which is composed of the **visceral, pedal,** and in front of the latter, the **brachial ganglia.** Connected with the sides of the cerebral mass are the two **optic nerves,** which widen out to form the large **optic ganglia,** and running forward from it are two small nerves which connect it with the **suprapharyngeal ganglia,** a small mass above the hinder end of the pharynx.   From these ganglia small nerves pass around the œsophagus to the pair of **subpharyngeal** ganglia. From the forward surface of the subœsophageal mass, *i.e.,* from the brachial ganglia, ten nerves pass off to the arms.   These may be seen on the inner surface of the head after the removal of the pharynx.   From the hinder surface of the visceral ganglia **pleural nerves** run to the **stellate ganglia** in the mantle. Trace these nerves from the stellate ganglia to their source.

   The pen.   Make a longitudinal slit in the mantle on the back of the animal and remove the pen; it will be seen to lie quite loosely in its sac.

Exercise 8. Draw the pen.

CHAPTER VI

TUNICATA

ASCIDIACEA

A SIMPLE ASCIDIAN (*Molgula*)

Ascidians are sessile, marine animals which live attached to rocks, seaweed, and other objects in the waters along our shores. Many ascidians are colonial animals; the young individuals, which arise by a process of budding, remaining attached to the parents. In a colony which is thus formed certain organs are often possessed in common, and a very intimate relation is established between its individual members. Molgula is noncolonial; it is usually found in clusters attached to rocks below low tide.

Molgula is a small saccular animal, an inch or less in length. Its outer covering is a thick, tough **tunic** or **test**, which is characterized by being partly composed of cellulose, a substance rarely met with in animals. The surface of the tunic is covered with numerous minute projections, among which sand and dirt lodge and cause the dirty appearance which characterizes it, except where it is in contact with that of other individuals.

The animal has two external body-openings, the **incurrent opening** or the mouth and the **excurrent opening**, each of which is at the end of a projection of the body-wall called a **siphon** and is fringed by short **tentacles**. The tentacles may, however, have been drawn into the openings and thus not be apparent. The incurrent siphon is at the anterior end of the body, the excurrent siphon represents the morphologically posterior end; the

135

portion of the body lying immediately between the two is the dorsal side ; the opposite side, which is very much longer and includes the surface of attachment, is the ventral side. A stream of water is drawn into the incurrent opening, bearing the minute organisms which constitute the animal's food and the air needed for respiration; through the excurrent opening the water is ejected, charged with fæcal matter and reproductive products.

Exercise 1. Make a sketch of the animal on a scale of 2 or 3; label the dorsal and ventral aspects and the siphons.

Beneath the tunic and in contact with it is the **mantle**, which is the remainder of the body-wall, the tunic being a highly modified **cuticula** protecting its outer surface. Remove the entire tunic. This may be easily done by snipping it with scissors and then pulling it off with forceps; it is not tightly joined with the mantle. The mantle will be seen to be a transparent structure through which the internal organs appear. Observe the white **muscle bands** in the mantle, especially the **transverse** and **longitudinal muscles** in the siphons by means of which they are extended and contracted. Note also the short **tentacles** at the incurrent and excurrent openings. Count those at each opening.

The digestive system. The most conspicuous internal organs are the cream-colored **genital glands** near the center of the body and the **alimentary canal.** The latter lies on the left side of the body, where it appears as an S-shaped structure which encloses the former. Place the body in water with the left side uppermost and the siphons away from you, and study the arrangement of the organs. The incurrent opening (at your left) will be seen to have more prominent tentacles than the excurrent opening. From the base of the incurrent siphon the large **pharynx**, the most voluminous organ of the body and the principal organ of respiration, will be seen extending to the lower side of the body. Note the six longitudinal ridges which

appear as light-colored bands in the pharyngeal wall. Find
and trace a white or cream-colored line extending in the mid-
ventral line from the base of the incurrent siphon to the opposite
side of the body. This is the **endostyle**; it is a ciliated and
glandular groove which lies between two folds in the mid-
ventral wall of the pharynx; it extends the length of that
structure and ends posteriorly near the opening of the pharynx
into the **œsophagus**. Find this point. The œsophagus is short
and communicates with the **stomach**, and these two divisions
form the lower and thicker limb of the S-shaped digestive
tract. The upper limb is formed by the **intestine**, which passes
to the base of the excurrent siphon, where it ends with the
anus. Find these organs.

The reproductive system. Molgula is hermaphroditic. The
sexual organs consist of a pair of large **hermaphroditic glands,** one
of which is seen on each of the lateral sides of the body.
A short duct runs from each gland to the base of the excurrent
siphon. On the left side the duct will be seen alongside the
posterior end of the intestine; find it.

The circulatory system. On the right side of the body beneath
the hermaphroditic gland will be seen the **heart** in its **pericardium.**
It is a muscular sac from each end of which proceeds a large
blood vessel. The vessel leaving the ventral end (at the
observer's right) is called the **cardio-branchial vessel**; it passes
along the mid-ventral side of the pharynx, beneath (external to)
the endostyle, and gives off branches which run transversely
along the pharyngeal wall. The vessel leaving the dorsal end of
the heart is called the **cardio-visceral**; it breaks up into numerous
branches, which ramify among the viscera and other parts of
the body. From the viscera the blood is collected again in
a vessel called the **viscero-branchial,** which passes along the mid-
dorsal pharyngeal wall and gives off **transverse branches.**

The heart of tunicates is peculiar in that its pulsations
change the direction of the flow of the blood alternately from

the cardio-branchial to the cardio-visceral vessels, and back again. The contraction of the heart is of a peristaltic nature; it passes from one end to the other of it for a short time; then after a short pause the contraction is renewed, the peristaltic motion beginning at the opposite end and driving the blood in the opposite direction.

The nervous system. About halfway between the two siphons, imbedded in the mantle beneath the dorsal surface of the animal, lies a small **ganglion** from which **nerves** radiate. No organs of special sense are present, except the tentacles and minute eye-spots at the incurrent and excurrent openings.

The excretory system. Beneath the heart is an elongated vesicular organ which is the single, unpaired **kidney**; it is ductless. Beneath the ganglion above mentioned is a small glandular organ called the **subneural gland**; it has a duct which communicates with the pharynx. The function of this gland is probably excretory; it is supposed to be homologous to the hypophysis of vertebrates.

Exercise 2. Make a drawing of the left side of the animal on a scale of from 4 to 6, showing all the internal organs which appear in that aspect. Label the dorsal and the ventral sides of the body and all the organs.

Exercise 3. Make a drawing of the right side of the animal showing all the organs which appear in that aspect.

Exercise 4. Make a drawing of the dorsal side showing the organs observed there.

The peribranchial chamber. Cut off the excurrent siphon at its base and with a needle or bristle probe the opening. The probe will pass into the large space between the mantle and the pharynx. This is the **peribranchial chamber**; it surrounds the pharynx on all sides, except in the mid-ventral line, and communicates with the outside water through the excurrent siphon.

It is not a part of the body-cavity, but has been formed by an infolding of the outer surface of the body. Into it, near the base of the excurrent siphon, the digestive and genital tracts discharge their products for removal with the current of respiratory water which streams out of that siphon.

The respiratory system. The principal respiratory organ is the pharynx, which communicates with the incurrent siphon by an opening fringed with a circular row of branched tentacles. Its walls are pierced by numerous slit-like, ciliated openings, called **stigmata**, through which the respiratory water streams from it into the peribranchial chamber. A current of water is thus maintained, which passes through the incurrent siphon into the pharynx, and thence through the stigmata into the peribranchial chamber, and out again at the excurrent siphon. The stigmata are vertical in position and are arranged in transverse rows, which extend across the pharyngeal wall, and are separated from one another by delicate **vertical bars**; the transverse rows have between them large **transverse bars**, and running longitudinally along the pharyngeal wall on each side are six large **longitudinal bars** or **ridges**, which are easily seen and have already been mentioned. Through all of these bars the blood circulates, being brought to them either by the cardio-branchial or the viscero-branchial blood vessels, and respiration is thus carried on.

Lay the animal with the left side uppermost. Slit open the incurrent siphon and the pharynx by inserting the point of fine scissors into the siphon and, after cutting its wall to its base, carrying the cut through the wall of the pharynx along the side of and parallel with the mid-ventral line to the posterior end of that organ. Lay the pharynx open. The twelve large longitudinal bars will be seen projecting into the pharyngeal lumen. Trace them throughout their entire extent. Find with the aid of a dissecting microscope or a hand lens the row of branched **tentacles** at the base of the incurrent siphon and count them.

In the mid-ventral line note the **endostyle**; notice also that it is a groove. Trace the endostyle forward to the base of the siphon. At its anterior end the endostyle is continuous with a ciliated ridge which encircles the anterior end of the pharynx and is called the **peripharyngeal ridge**. This ridge is itself continuous, on the dorsal side of the animal, *i.e.*, on the side opposite to the endostyle, with a ciliated longitudinal ridge called the **dorsal lamina**, which passes along the mid-dorsal line to the opening of the œsophagus at the posterior end of the pharynx. Trace the peripharyngeal ridge and the dorsal lamina.

These organs aid in the ingestion of the animal's food, which consists of minute organisms and particles of organic matter. The endostyle is a glandular and ciliated groove; the gland-cells secrete a viscid substance which catches the food particles; the cilia create a current which drives them towards the ante-rior end. Here they meet a current created by the cilia of the peripharyngeal ridges which takes them around the pharyngeal wall to the dorsal lamina, along which they are driven poste-riorly to the opening of the œsophagus.

Between the two siphons note the **ganglion** and, beneath it, the **subneural gland**.

Exercise 5. Make a semidiagrammatic drawing showing the structures which appear in connection with the pharyn-geal wall.

Exercise 6. Make a large diagram of Molgula and show the rela-tive positions of the different organs; label all.

CHAPTER VII

ECHINODERMATA

ASTEROIDEA

A STARFISH

Several species of starfishes are common along our coasts, the most familiar being *Asterias vulgaris*, the common New England form, which is found along the entire Atlantic coast, and *Asterias forbsii*, which is found south of Cape Cod. They are remarkably sluggish creatures which live on the sea bottom, moving slowly, often in large numbers, from place to place and feeding on the various mollusks which come in their way.

Two specimens will be needed for this dissection, a dried one for the study of the hard parts, and one that is fresh or has been preserved in formalin or alcohol for the study of the internal and other soft parts. To prepare a dried starfish the live animal should be placed in fresh water for half an hour. It should then be placed in alcohol for an hour, and then dried thoroughly. If only preserved material be at hand the animals may be simply dried. The fresh water and alcohol expand the body-wall of the animals and prevent it from collapsing after death.

Study the external characters of a fresh or a preserved specimen. Observe the color and the flattened radiate body-form. The body is composed of a **central disc** from which radiate five **arms or rays**. All of these rays are normally of equal length. Specimens are often found, however, in which the length of the rays is unequal. This is due to the fact that starfishes often lose one or more of their rays by accident; the missing member is soon replaced by a new ray, but while it is growing out it will be shorter than the others. The spaces between the rays are called **interrays**. In the center of the under surface of the disc is the **mouth**; hence this surface of the animal is called the **oral surface**. Its upper surface is called the **aboral surface.**

In the aboral surface of the disc notice the red **madreporic plate** (in preserved specimens it may have lost its color and be white). Examine it on the dried specimen with the aid of a hand lens or the low power of a compound microscope and notice its porous structure. In the aboral surface is also the **anus**; it is a very small opening and will be difficult or impossible to see in the specimens at hand. Note the short fixed **spines** covering the entire aboral surface. Each one is a part of a small **calcareous plate** buried beneath the integument. The entire body-wall of the animal is made up largely of these plates, which give it its stiffness. The plates are not, however, connected with one another except by muscles and connective tissue, and the animal's arms are, consequently, flexible and freely movable. Demonstrate this fact with your specimen. In the dried animal this flexibility no longer appears, as the entire body-wall has been rendered rigid by the drying. In the soft places between the plates note the delicate tubular projections of the integument; they are the contractile **papulæ**, and are organs of respiration and excretion and possibly also of sensation. With the aid of a hand lens find, around the base of each spine, the **pedicellariæ**; these are minute pincer-like organs of somewhat uncertain function, but which probably aid in keeping the surface of the animal free from particles of dirt and from minute organisms which might be harmful.

The two arms which enclose the madreporic plate between **their bases are called the bivium**; the remaining three, the trivium. How can a plane be passed through the body so as to divide it into two symmetrical halves?

Exercise 1. Make a life-size drawing of the aboral aspect of the animal and label all the features observed.

On the oral surface observe the deep groove which extends from the mouth along each arm to its tip. This is the **ambulacral groove.** Observe the two rows of movable spines which fringe each side of the groove; also the five pairs of movable spines which surround the mouth. Separate these spines and observe the mouth surrounded by a circular membrane, the **peristome.**

From the sides of each ambulacral groove two zigzag rows of soft tentacles project. These are the **ambulacral feet**; they are muscular tubes with sucker discs at their ends and are the organs of locomotion. Scrape the feet from a portion of the groove and examine its sides; note the slender, transverse, **calcareous plates** which form it, and the round openings between them, called the **ambulacral pores**, through which branches from the feet project into the body-cavity. Note the zigzag nature of each of the two rows of these pores. Notice also the delicate cord which extends along the median line of the groove; it is the main **nerve** of the arm; it proceeds from a **nerve ring** in the central disc to the tip of the arm. Follow it to the tip and note the red pigment spot with which it ends. This is the **eye.** In preserved specimens the pigment may have lost its color.

Exercise 2. Make a life-size drawing of the oral aspect of the animal and label all of these features.

Scrape off several pedicellariæ, mount them on a slide, and examine them under a compound microscope. By pressing on the cover-glass with a needle, the jaws can be made to open and shut; try it.

Exercise 3. Draw a pedicellaria on a large scale.

Cut off an arm of the dried specimen, and also a bivial arm of the fresh one, and examine the cut surface of each. The edge of the calcareous plates will be seen, as well as the spaces between them. Notice the slender plates which form the sides of the ambulacral groove; also just beneath the median ridge, in the upper part of the apex of the groove, a minute opening. This opening is the **radial canal**, which extends the length of the arm; its function will be explained when the ambulacral system is described. If a portion of the arm be soaked for a short time in a strong solution of warm caustic potash, the soft parts will be destroyed and the plates will be seen more distinctly. Care should be taken not to allow the potash to act too long or the arm will fall to pieces.

Exercise 4. Make a sketch of the cut edge of the arm on a scale of 3, showing the edges of the plates and the radial canal.

Cut off the aboral wall of the severed arm of the dried specimen and scrape away the remains of the internal organs and the ambulacral feet. Study the inner surface of the ambulacral groove. Note the two rows of slender **transverse plates** which form the sides of the groove, and on each side between every two plates, the minute **ambulacral pore**.

Exercise 5. Make a drawing on a scale of 3 of the inner surface of the ambulacral groove, showing the plates and the pores.

Cut off the aboral wall of the central disc of the dried specimen, scrape away the remains of the internal organs, and study the arrangement of the plates in the inner surface of the oral body-wall. Note the circular **mouth** protected by converging spines, also the membranous **peristome**. Observe the convergence of the five arms about the peristome; also the **interradial partitions** which separate the base of the arms.

Exercise 6. Make a life-size drawing showing these features.

Internal anatomy. Remove the entire aboral body-wall from the trivium and the central disc of the fresh or preserved specimen, with the exception of the madreporic plate which must not be removed, being very careful not to injure the organs beneath. Study the internal organs and observe the following systems:

The digestive system. Observe the large sac-like **stomach,** which almost fills the central disc. Its walls are much folded, and five short, bag-like pouches extend from it into the five arms. When the animal feeds the stomach is everted and thrust out through the mouth and about its prey. It is drawn in again by means of five pairs of **retractor muscles,** which connect the stomach pouches with the inner surface of the ambulacral grooves. Find the pair of retractors belonging to each stomach pouch. Communicating with the aboral portion of the stomach are five large radial digestive glands, which are usually called **livers.** Each gland almost fills an arm ; it is made up of two main trunks, from which project numerous side branches ; the two ducts leading from the two trunks in each arm unite to form a single duct which passes to the stomach. Each trunk is suspended from the aboral wall of the arm by two **mesenteries.** Find the mesenteries in one of the bivial arms. Study the structure of the livers. The stomach is connected with the mouth by a short **œsophagus,** and from its upper surface a short slender **intestine** passes to the anus. Connected with the intestine is a small branched diverticulum, the **intestinal cæcum.** The intestine, together with its cæcum, may have been removed when the aboral body-wall was taken off. If this be the case look for them on the portion of the aboral wall which was taken off and notice also the position of the anus.

The reproductive system. The sexes of the starfishes are separate. The sexual organs are branched **glandular organs,** ten in number, which lie in the proximal portion of the rays and open to the outside through minute pores in the aboral walls of the interrays. Two glands will be found in each ray extending from the base of the ray toward its tip. The actual size of these organs depends entirely upon the sexual condition of the animal. In young or immature animals they may be no more than half an inch long or less, while in reproducing animals they may extend almost to the tip of the ray. The **testis** of the male and the **ovary** of the female animal do not differ from each other in general appearance. In the mature female, however, the ovaries have a light-yellow color, while in the mature male the testes are white and are less voluminous than the ovaries.

Exercise 7. Make a semidiagrammatic drawing of the animal showing the details of the digestive and reproductive systems; label all.

Remove the stomach and the reproductive organs from the body, taking care not to injure the sinuous stone canal which is at one side of the former.

The ambulacral system. This is the most characteristic system of organs in the Echinodermata. In the starfish it consists of the following organs: a circular canal, called the **ring canal**, surrounding the mouth; connected with this canal are nine minute lobated sacs called the **racemose** or **Tiedemann's vesicles**, two being located in each interray except the one in which is the stone canal, where but one is present; five **radial canals**, which pass from the ring canal along the median line of the ambulacral grooves to the tips of the arms; the **ambulacral feet**, which are connected with the radial canals by short **branch canals**, and also project through the ambulacral pores into the body-cavity, where they expand to form small sacs called **ampullæ**; a sinuous canal, called the **stone canal**, which connects the ring canal with the madreporic plate; the **madreporic plate**, a porous plate by means of which the entire system is placed in communication with the outside sea water.

In studying this system find first the madreporic plate and the stone canal, and trace the latter from the madreporic plate to the ring canal. Remove the spines which project over the peristome and find the ring canal. It is a delicate tube, of about the diameter of a needle, which surrounds the mouth, running around the base of the arms at the point where the peristome joins them; it is thus, like the radial canals, outside the body-cavity. Remove some of the ambulacral feet from a ray, and find again the delicate radial canal which lies along the middle of the ambulacral groove. Trace it to the ring canal. Cut the aboral body-wall from one of the bivial rays, remove the liver, and observe the ampullæ. Press them and notice that the feet are thereby extended.

The ambulacral system will be seen to be a system of tubes extending throughout the body and in communication with the sea water. They are filled with a fluid which is not, however, pure sea water, but is rather a watery serum in which float amœboid cells. This fluid is driven into the tube-like ambulacral feet, which thereby acquire rigidity and are extended. The system is the locomotory system of the animal. It moves by extending the feet, attaching the sucker discs at their ends to some stationary object, and then drawing them in. The animal is thus able to pull itself slowly along. The ambulacral system possibly also exercises excretory and respiratory functions.

Exercise 8. Draw a diagram of the ambulacral system.

There are no special respiratory and excretory organs. These functions are exercised by the papulæ and possibly the ambulacral feet.

The **nervous system** consists of a circumoral **nerve ring**, which lies just beneath the ambulacral ring canal, and five **radial nerves**, which proceed from it along the median line of the ambulacral grooves to the tips of the arms. Each radial nerve ends with a **pigment eye**. There are no other organs of special sense. The main nerves of the starfish do not lie within the body-cavity, but in the integument, and can thus be seen from the outside. There are, however, in addition to these nerves, other less important ones which are internal. We have already observed the radial nerves in the median line of the ambulacral grooves; the ring nerve can also be seen as a slight ridge just beneath the ring canal.

Exercise 9. Draw a diagram representing the nervous system.

The **circulatory system** consists of a very complicated system of tubes and spaces, filled with a blood fluid, none of which can be seen in a dissection, except an organ usually called the **heart** or **axial sinus**. This is a tubular sac which will be found beside the stone canal; within it is an elongated glandular organ called the **axial organ** or **ovoid gland**.

Exercise 10. Draw a diagram representing a vertical section of the animal passing through the madreporic plate and the anterior ray (*i.e.*, the middle trivial ray).

ECHINOIDEA

A SEA URCHIN

Several species of sea urchins occur along the Atlantic coast, the most familiar being Arbacia, the dark-colored urchin, and Strongylocentrotus, the green urchin, the former having a more southerly distribution than the latter. The animals live on the sea bottom or on rocks, usually in companies, and move slowly about from place to place, using not only the ambulacral feet, but often the spines as well, as organs of locomotion. They feed partly upon small animals and partly upon organic substances present in the sand and mud, which they pass through the intestine.

Two specimens will be needed for this dissection, a dried one for the study of the hard parts and a fresh or preserved one for the study of the internal organs.

Observe the radiate spheroidal body entirely covered with movable spines. Look among the spines and find the **ambulacral feet.** These can be extended in life beyond the spines and are employed by the animal as organs of locomotion. Note the five **ambulacral areas** (those containing the feet), and between them the five **interambulacral areas.** The flattened surface is the under or **oral surface,** on which the animal moves; the rounded surface is the **aboral.** It will be seen that the aboral side of the sea urchin bears ambulacral feet, whereas in the starfish the oral side alone bears them.

In the center of the oral surface, observe the **mouth** and the five calcareous **teeth** which project from it. Surrounding the mouth is a membrane which fills the space between the edges of the shell and is called the **peristome.** Notice the ten short ambulacral suckers which surround the mouth, and near them the five groups of pincer-like pedicellariæ. Observe the long slender stalks of these organs. Near the margin of the peristome are five groups of ambulacral feet.

Exercise 1. Make a drawing of the oral surface on a scale of 2.

With forceps remove some of the pedicellariæ, mount them, and study them under the microscope. Note the three minute jaws and the long stalk. Press on the cover-glass and cause the jaws to open and shut.

Exercise 2. Make a drawing of a pedicellaria.

Study the structure and method of articulation of the spines. Pull off several and notice their ball and socket joint, also the delicate muscles by which they are moved. Notice the fluting of the shaft.

Exercise 3. Make a semidiagrammatic drawing of a spine on a large scale showing the articulation and the muscles.

Remove the spines from the dried specimen and thoroughly clean the shell. This is accomplished the most effectually by placing it in a strong solution of warm caustic potash for a short time. Great care should be taken, however, not to leave it in the solution too long or it will fall to pieces. Study the aboral side of the shell. Observe the rows of tubercles on which the spines have articulated, also the bands of minute holes, the ambulacral pores, through which the ambulacral feet have projected. There are ten of these bands arranged in pairs, and each pair represents an ambulacral area or a ray. Between the five rays are the five interambulacral areas or the interrays, which are somewhat broader than the rays; count the rays and the interrays.

The center of the aboral surface is free from spines and is made up of several small plates. It is called the periproct and contains in its center the minute **anus**. Surrounding the periproct are ten plates, which also bear no spines. Five of these, which are larger than the others, are situated at the ends of the interrays and are pierced each by a small hole. These plates are called the **genital plates**, and the holes are the external openings of the genital organs. One of the genital plates is larger than the others and is porous; it is the **madreporic plate**. The five smaller of the ten plates which surround the periproct are situated at the ends of the rays. They are called the **ocular plates**. Each is pierced by one or two holes, through which project minute pigmented tentacles. Notice that each ray and each interray is made up of two rows of plates, so that there are twenty rows of plates altogether. As in the starfish, the two rays between which the madreporic plate lies are called the **bivium**, the other three, the **trivium**.

Exercise 4. Make a drawing of the aboral side of the shell, with the spines removed, on a scale of 2, showing accurately the boundaries of all the plates. Label the rays, interrays, and all the other parts observed.

The internal organs. Place a fresh or preserved sea urchin in a pan of water. Carefully cut away the peristome with scissors and remove the shell of the oral body-wall on one side of the peristome without disturbing the organs within. Observe the following systems of organs:

The digestive system. This is quite different from the same system in the starfish. The mouth opens into the œsophagus, which passes through the center of the large cone-shaped **dentary apparatus**, which is also, because of its shape, called **Aristotle's lantern**. This is a complicated structure consisting of a number of calcareous plates and muscles which project from the mouth into the body-cavity. Study its muscular attachment with the shell. Note the **protractor muscles** which pass from its upper end to the oral body-wall, by means of which the apparatus can be thrust down and partly out of the mouth; also the **retractor muscles** which pass from the lower part of it to the tall inner projections of the shell.

Exercise 5. Draw a diagram showing the dentary apparatus in the body and its muscular attachment to the shell.

The œsophagus, after leaving the dentary apparatus, passes to the elongated **stomach**; this lies close to the body-wall, to which it is attached by means of a **mesentery**. Carefully follow the stomach, breaking away the wall if necessary, as it winds around the inner surface of the shell. From the stomach a short **intestine** passes to the **anus**. In making this dissection, keep the animal immersed in clear water; remove as little of the shell as possible, and do not remove any of the organs from the body.

The **genital system** is similar to that in the starfish. The sexes are separate and the **sexual glands** of the male and female are alike in appearance. They consist of five radial, granular masses, which lie in the upper part of the body-cavity, each mass communicating with the outside through one of the genital pores. The actual extent of the sexual glands depends upon the sexual condition of the animal. During the breeding season, in the summer, they may almost fill the body-cavity.

Exercise 6. Make a diagrammatic drawing of the digestive and the reproductive systems and label all their parts.

Remove the dentary apparatus from the body and examine it carefully. It is made up principally of five triangular plates called **alveoli**, the lower ends of which bear the **teeth**. The alveoli are bound together by short muscles. The base of the dentary apparatus is made up of a complicated system of smaller plates.

Exercise 7. Make a drawing of the dentary apparatus.

The **ambulacral system** is similar to that of the starfish. A **ring canal** surrounds the œsophagus just inside the inner end of the dentary apparatus and is connected with the madreporic plate by means of the **stone canal**. This organ is a small tube which lies in contact with the œsophagus and also, in the neighborhood of the aboral body-wall, with the intestine. From the ring canal five **radial canals** pass along the median lines of the rays to their aboral ends, sending off branches to the **ambulacral feet**. The entire system of tubes, except the ambulacral feet, is within the body-cavity, instead of outside of it, as in the starfish. Look on the inner surface of a ray for the radial canal. On each side of it observe the row of small vesicles, the **ampullæ**, the reservoirs of the ambulacral feet. Determine the exact relation of the ampullæ to the feet, and of both to the ambulacral pores in the shell. It will be seen that there are two rows of these latter on each side of the radial canal. Through one of these rows the branch canals pass from the radial canal to the ambulacral feet on the outside of the shell; through the other row projections of the feet pass back into the body-cavity, where they expand to form the ampullæ. There is thus a single row of feet on each side of the radial canal in each ray.

The ambulacral system will be seen to consist of a system of tubes extending throughout the body, and communicating with the sea water through the madreporic plate. It is filled with a fluid which, as in the starfish, is not pure sea water, but is rather a watery serum in which float amœboid cells. This fluid can be driven into the ambulacral feet, which acquire rigidity and are thereby extended. The animal moves by extending the feet, attaching the sucker discs at their ends to some stationary object, and then drawing them in; it is able thus to pull itself slowly along. Some sea urchins with long spines also move on the tips of their oral spines as on stilts.

Exercise 8. Draw a diagram representing the ambulacral system.

The respiratory and excretory systems are not represented by special organs. The ambulacral feet may possibly perform both functions.

The **nervous system** cannot be studied in a dissection. It is essentially like that of the starfish. From the ring nerve around the œsophagus five radial nerves pass along the median line of the rays to the ocular plates, where each terminates in a tentacle. The entire system is within the body-cavity instead of a portion of it being outside as in the starfish.

The **circulatory system** cannot be seen in a dissection. As in the starfish it consists of a complicated system of tubes and spaces which contain a colorless blood fluid. An organ of problematical function called the **heart** or **axial sinus** is found attached to the upper end of the stone canal; within it is a glandular organ called the **ovoid gland** or **axial organ**.

HOLOTHURIOIDEA

A HOLOTHURIAN OR SEA CUCUMBER

Two common species of holothurians which live in the shallow water of the New England coast are suitable for this dissection — *Thyone briareus* and *Cucumaria frondosa*, the former being common in Vineyard and Long Island Sounds and the latter on the Maine coast. The latter is the larger of the two species, being from 10 to 25 cm. or more long and about half as thick, and differs from Thyone, among other things, in having ambulacral feet in the radial areas alone; it possesses thus five broad bands of these appendages, the interradial areas being smooth. Thyone is from 8 to about 20 cm. long and 5 cm. thick, and is covered all over with ambulacral feet, the five broad radial bands meeting in the interradial areas. It will be used as the basis of this dissection.

Observe the form and color of the **body**. Note that the upper and lower sides are distinctly differentiated, the former having fewer feet than the latter. In the center of the forward end is the **mouth**, surrounded by the ten branched **tentacles** by means of which the animal collects the minute organisms which constitute its food. Above the mouth and between the bases of two tentacles is the **genital pore**. In the center of the hinder end is the **anus**, surrounded by five **anal teeth**. The main longitudinal axis of the body (that which joins the mouth and the anus) will thus be seen to be very long, in sharp contrast to that of the starfish and sea urchin, in which it is much shorter. In consequence of this feature the body is elongate and more or less worm-like, and the animal does not rest on the oral surface but on its side.

Exercise 1. Draw a side view of the animal and label carefully all of the features observed.

With strong scissors cut the body open by a longitudinal slit along the underside. Place it in a pan of water and with large pins pin down the two sides of the body wall on the right and left. Without cutting any of them, study the internal organs.

Note the spacious body cavity and the long, coiled intestine which partly fills it. At the front end of the body and just behind the tentacles is the conspicuous **calcareous ring**, a more or less rigid cylinder containing five radial and five interradial plates, which surrounds the œsophagus. Projecting from the hinder end of the calcareous ring will be seen two large cylindrical sacs, the **Polian vesicles**, and inserted in the ring are five prominent **retractor muscles**, by means of which the tentacles and the forward end of the body can be invaginated. Note the position of the **genital gland** and its **duct**. The gland is a thick bunch of slender filaments, in the forward part of the body cavity, joined with the body wall by a mesentery, which converge to the hinder end of the duct. This is a slender tube which passes forward on the upper side of the body cavity and opens to the outside between two tentacles on the upper side of the body. The sexes are separate; they are alike, however, in appearance. Note the **respiratory trees**, the profusely branched organs which extend from the rectum throughout the entire length of the body cavity.

Study the course of the **digestive tract**. The **œsophagus** passes through the calcareous ring to the thick-walled, muscular **stomach**, which lies directly behind it. From the hinder end of the stomach the long, thin-walled **intestine** passes, with many loops and turns, to the short, wide **rectum** at the hinder end of the body. Trace the intestine carefully, but without cutting it, throughout its entire course and note the three mesenteries by which it is attached to the body wall.

Note the thick muscular walls of the rectum and the **muscle strands** which join it with the body wall. The two branching respiratory trees spring from the forward end of the rectum and extend through the body cavity to its forward end. Observe carefully their extent. It is by the periodic dilation and contraction of the muscular walls of the rectum that water is alternately taken into the rectum and the respiratory trees and expelled from them, and respiration thus carried on.

Exercise 2. Draw a diagrammatic view of the opened animal on a large scale, showing the organs observed. Carefully label all.

Cut the œsophagus just back of the calcareous ring and remove the digestive tract with the respiratory trees and the genital gland from the body. The five **retractor muscles** of the calcareous ring may be followed to the body wall, where they will be seen to join five **longitudinal muscles** which extend along the inner surface of the body wall the length of the body. These muscles mark the five radial areas of the body. Note the **circular muscles** which also lie on the inner surface of the body wall.

Study the **ambulacral system**. The **ring canal**, which is often difficult to see, surrounds the œsophagus at the hinder end of the calcareous ring. The **Polian vesicles**, two elongated sacs which have already been noted, extend from it into the body cavity. They secrete and store the lymphatic fluid which fills the ambulacral canals. The **stone canal**, which also extends from the ring canal, is a slender tube, and the **madreporic plate** with which it ends lies in the body cavity and not at the surface of the body. It is probably a rudimentary structure. The five **radial canals** extend from the ring canal along the radial areas of the body and beneath the five radial **muscle bands** to the hinder end of the body. The **ambulacral feet** extend from the radial canals through the body wall and are seen thickly studding the outer surface of the body. The numerous **ampullæ** extend into the body cavity and will be seen on the inner surface of the body wall. The **tentacles** are also a part of the ambulacral system, the canals which supply them with the ambulacral fluid springing from the radial canals.

Exercise 3. Draw a diagram of the ambulacral system.

Exercise 4. Draw a diagram representing an ideal cross section through the middle of the body.

The **calcareous spicules** and **plates** which form a part of the body wall and are so characteristic of the echinoderms as a group are less conspicuous in holothurians than in any of the other classes of the phylum, and will not be studied in this dissection. They are of minute size, although definite in shape in the various species, and do not form a connected skeleton as in the starfish and the sea urchin. Pedicellariæ are absent.

CHAPTER VIII

CNIDARIA

HYDROZOA

A FRESHWATER POLYP. HYDRA

This small animal is of general although sporadic distribution throughout the country. It frequents ponds and streams abounding in plant life, and is best obtained by collecting water together with water plants, sticks, and other objects from several such places, and allowing it to stand in glass jars. The polyps will, if present, be seen, after an interval of some time, attached to the stems or leaves of the plants, or to the sides of the jar. They may be kept indefinitely in aquaria of this sort and will usually multiply rapidly.

Hydra is a slender, tubular animal from one-eighth to one-half an inch in length; it attaches itself by one end to some stationary object and projects pendant in the water; at the other end is the **mouth**, surrounded by from four to eleven long thread-like **tentacles**. It does not attach itself permanently to one place, but can crawl about or swim slowly through the water. Its food consists largely of small crustaceans which it kills or paralyzes with the peculiar stinging organs called **nettle cells**, located principally in the tentacles.

Study the animals first, if possible, without disturbing them and with the aid of a hand lens. Note the extreme contractility of the body. Look for individuals with distended bodies. These have just swallowed prey. Look for budding individuals; budding is one method of reproduction. Observe the color of the animal. Two species of the hydra are common in this country — the green hydra and the brown hydra, the latter of which has much longer tentacles than the former.

Detach a polyp by means of a pipette, place it in a watch-glass of water, and study it under a microscope.

Exercise 1. Draw several outlines of the animal on a large scale, showing different shapes and positions it can assume. Notice the mouth in the midst of the tentacles.

Place a polyp on a slide in water and cover it with a cover-glass, but support the corners of the latter with wax to avoid crushing the animal. Study its structure under the microscope. We see that the type of structure is radial and not bilaterally symmetrical; that the animal is tubular in shape, and that its internal cavity opens to the outside through the mouth; that the **mouth** is a small opening at the end of the conical, terminal portion of the body, called the **hypostome**, at the base of which are the tentacles. The internal cavity is called the **gastro-vascular space** and is the common digestive and circulatory cavity of the animal. The end by which the animal is attached is called the **foot**. It is an adhesive disc containing gland-cells which produce a sticky secretion.

Examine the finer structure under a high power of the microscope. The body-wall is made up of two layers of cells, the outer **ectoderm** and the inner **entoderm**. The cells of the latter are much longer and more irregular than those of the ectoderm; their inner surfaces are amœboid and also often flagellate. Imbedded in the entoderm cells are the **chromato-phores**, the bodies which give color to the animals. In the green hydra these are chlorophyll bodies; in the brown hydra they consist of a substance similar to chlorophyll. In both cases they are probably single-celled algæ living symbioti-cally with the polyp. Between these two cell layers is a thin non-cellular one called the **supporting layer**. The tentacle is a hollow projection of the body-wall and has the same structure.

Among the ectoderm cells of the distal portion of the body, and especially of the tentacles, notice the oval, highly refractive stinging organs or **nematocysts**. Each one consists of a spiral thread-like tube, with several barbs at its base, which lies coiled within the cavity of a cell called a **cnidoblast**. The cavity is filled with a poisonous fluid; its walls form an ovoid sac, of which the tube is the very much elongated and invaginated outer end. A minute spine projects beyond the free surface of the cnidoblast into the water. When the surface of the ectoderm is irritated the tube is evaginated and shot violently out, and the poisonous fluid contained in the cavity of the nematocyst is injected into any animal that may be struck.

Exercise 2. Make a large semidiagrammatic drawing of the animal showing the details of its structure; label all carefully.

Methods of reproduction. Hydra reproduces both sexually and asexually. Well-fed polyps will soon begin to bud off new individuals. The bud makes its appearance first as a projection of the body-wall, and soon becomes a distinct branch. Tentacles and a mouth make their appearance at the extremity of the branch, and the young polyp is complete. It remains attached to the parent for a while, then detaches itself and begins an independent life.

Besides this asexual method of reproduction, which may go on as long as the animal is well fed and vigorous, reproduction by sexual methods also occurs at more or less irregular intervals. **Sexual organs** appear in the form of projections of the ectoderm of the body-wall. Two classes of these projections appear, smaller ones, which are **testes**, and larger ones, which are **ovaries**. The former of these organs, which lie near the distal end of the animal, produce spermatozoa; the latter are near the proximal end, and each produces a single large ovum. The green hydra is a hermaphroditic animal, having both testes and ovary when sexually active, while the brown hydra is diœcious, being either male or female.

Exercise 3. Find a polyp with reproductive organs and make a drawing of it.

Special respiratory, excretory, digestive, and circulatory organs are absent in the hydra. **Respiration** and **excretion** are carried on through the surface of the body-wall. **Digestion, circulation,** and **absorption** go on within the gastro-vascular space. The animal's prey is caught in the water with the tentacles, which sting it into insensibility, and then swallowed into the gastro-vascular space. The entoderm cells, which line this cavity, extend their very plastic inner ends toward or about it, and some of them secrete a digestive fluid. The object is then digested and mingles with the water present in the gastro-vascular space, while by the vibrations of the flagella currents are produced which cause this fluid to circulate throughout all the parts of it.

Distinct muscular and nervous systems are absent in the hydra. Delicate **muscle fibers** are present, however, in the form of long parallel projections of the inner ends of ectoderm cells. Nervous elements are also present in the form of isolated **ganglionic cells** situated in the ectoderm, which send delicate projections to the muscle fibers and to the nematocysts. Both muscle fibers and ganglion cells are present throughout the animal's body but are most numerous in the tentacles. There are no organs of special sense.

HYDROZOA

A TUBULARIAN HYDROMEDUSAN (*Pennaria* or *Bougainvillea*)

These are marine animals, and among the commonest hydro-medusans along our coast. As is characteristic of the group to which they belong, they exhibit the phenomenon of **alternation of generations.** Two generations of individuals, a sexual and an asexual, alternate with each other. The latter is called the **hydroid generation**; the animal in this stage is sessile and colonial, and produces by budding, *i.e.*, by asexual methods, the sexual generation. This latter is called the **medusoid genera-tion**; in it the animal either remains attached to the hydroid colony (Pennaria) and is then called a **sporosac,** or separates itself (Pennaria, Bougainvillea) and becomes a free-swimming jelly-fish, which is called a **medusa**; in either case the medusoid produces by sexual methods embryos which attach themselves to fixed objects and develop into the hydroid generation.

The hydroid stage in Pennaria.[1] In this stage these animals form branching colonies of polyps, which are attached to the rocks in shallow water. The colonies are several inches in length, and are found in thick clusters which often cover the rocks over small spaces; their color is a delicate pink.

Place a small portion of a colony in a watch-glass of water or alcohol, and study it under the microscope. Observe the main stem of the colony and its branches, also the position of the **polyps** on the branches. Note carefully the differences in size between the different polyps. Which is the largest polyp? Study the method of branching. Has the colony a

---

[1] Bougainvillea or any other tubularian can also be used, with slight changes in the directions.

main axial stem? If it has, which is the oldest polyp? Which branch is the oldest? Which is the second oldest polyp? In Bougainvillea, besides the polyps, the medusoids also grow out from the stem.

The stem of the colony together with the branches is called the **hydrocaulus**; the root-like projections by which it is attached at its base are the **hydrorhiza**; the polyps are called **hydranths**.

Exercise 1. Draw a diagram representing the method of branching and showing the arrangement of the polyps on the stem.

Mount a small branch of the colony on a slide in water or dilute glycerine, and study a number of hydranths of all sizes. We note the radial type of structure and the tubular body, the internal cavity of which opens to the outside through the small terminal **mouth**. The stem also and its branches have an internal cavity which is a continuation of that of the hydranth. The cavity of the hydranth and the stem is called the **gastro-vascular space,** and is the common digestive and circulatory cavity of the animal. Notice the ringed constrictions in various parts of the stem, especially near the polyps; they give the stem strength and flexibility.

Study the two kinds of **tentacles**, the whorl of larger ones around the base of the hydranth and the shorter ones on the body of it. In Bougainvillea the basal tentacles alone are present. Count the basal tentacles. Count the short tentacles on a large and then on a small hydranth. The larger hydranth will be found to have more than the smaller one. Notice the small **hydranth buds** on the side of the stem; find one whose tentacles have not yet developed. Some of the hydranths will be seen to have large ovoid projections of very variable size on their sides. These are the **medusoid buds,** which become either free-swimming medusæ or sessile sporosacs. In Bougainvillea the medusoid buds do not appear on the hydranths but on the stem; they become free-swimming medusæ.

Exercise 2. Make a semidiagrammatic sketch of a large hydranth and a portion of the stem on a large scale; label the different parts.

Exercise 3. Make a sketch on a large scale of the smallest hydranth or hydranth bud you can find.

Mount a hydranth and a part of the stem on a slide in dilute glycerine, and study their finer structure. First study the structure of a tentacle. It is not hollow as is the tentacle of Hydra, but is made up of an axis consisting of a single row of large **entoderm cells** around which is a layer of **small ectoderm cells**. Between these two cell layers is the delicate non-cellular **supporting layer**. Find the highly refractive stinging organs or **nematocysts** at the end of the tentacle. These are the organs with which the animal kills its prey. Each one consists of a spiral thread-like tube, with several barbs at its base, which lies coiled within the cavity of a cell called the **cnidoblast**. The cavity is filled with a poisonous fluid; its walls form an ovoid sac, of which the tube is the very much elongated and invaginated outer end. A minute spine projects beyond the free surface of the cnidoblast into the water. When the surface of the ectoderm is irritated the tube is evaginated and shot violently out, and the poisonous fluid contained in the cavity of the nematocyst is injected into any animal that may be struck. Look for nematocysts which have discharged their spiral threads.

Exercise 4. Draw the distal portion of a tentacle showing its cellular structure; show the nematocysts at the end, including several which have been discharged.

Study the structure of the wall of the hydranth. It is made up of an outer **ectoderm** and a much thicker inner **entoderm**, each

composed of a single layer of cells; the inner ends of the entoderm cells are amœboid and often flagellate, the function of the
flagella being to maintain in circulation the fluids in the gastro-
vascular space; between these two layers is the thin non-cellular **supporting layer**. Study the structure of the stem. Observe
its central cavity, which is a part of the gastro-vascular space,
and the three layers just mentioned. In live specimens notice
the action of the flagella. Notice the cuticula which covers the
outer surface; it is not found in the hydranth. This cuticula
is called the **perisarc**. It is a supporting structure and gives
the colony rigidity.

Exercise 5. Make a drawing showing the cellular structure
        of the wall of the hydranth and of the stalk; carefully
        label all.

Study the structure of the **medusoid buds**. Notice the difference between the younger and the older buds. Compare the
young buds with the hydranth buds; compare the oldest
medusoid bud with the structure of the medusa, as described
later on.

Exercise 6. Draw the oldest and the youngest buds identifying
        as many of the parts as possible.

Special respiratory, excretory, digestive, and circulatory
organs are not present in the hydroid. **Respiration** and **excretion**
are carried on through the surface of the body-wall. **Digestion**,
circulation, and **absorption** go on within the gastro-vascular space.
The polyps feed upon small swimming animals, which they kill
or stun with their nematocysts, and then swallow into the
gastro-vascular space. Digestion goes on within this space
and waste matters are ejected from the mouth. The products
of digestion mingle with the water present in the gastro-vascular space and circulate throughout the colony, the internal
cavities of all the individual polyps of a colony being in

communication with one another. When a colony is well fed it grows rapidly, new hydranths bud out from the stalk, and medusoid buds appear and grow into medusæ. The hydranths are frequently destroyed by frost, or by the beating of waves, or by fishes, but new ones quickly grow in their places.

The medusoid stage. The medusoids of tubularian hydromedusans are either sessile **sporosacs** or free-swimming **medusæ**. Pennaria produces both kinds. During a greater part and in some cases the whole of the year they are sporosacs, but usually during the middle and last of the summer they detach themselves from the hydroid and become medusæ. Both conditions may be found in the same colony and at the same time. The medusoids of Bougainvillea are always free-swimming and with other medusæ will be found in the surface waters of the ocean. They may be easily obtained by placing the live hydroid colony in a small glass of sea water; the medusæ will be found swimming about in the water.

Place several medusæ of Bougainvillea or of any other tubularian hydromedusan in a watch-glass of sea water or, if they are preserved specimens, in alcohol. If they are alive, observe the swimming motions. Note the radiate type of structure. The body is bell-shaped, having an outer convex and an inner concave side, and on its margin are **tentacles** (in the medusa of Pennaria they are rudimentary). The convex side is called the **exumbrella** or **aboral side,** and the concave, the **subumbrella** or **oral side.** In the center of the latter is the proboscis-like projection called the **manubrium,** at the distal end of which is the **mouth,** surrounded by short **oral tentacles.** The mouth opens into the **gastro-vascular space,** which comprises the space within the manubrium and also a system of canals in the bell-shaped body. These canals consist of four **radial tubes,** which extend from the base of the manubrium to the periphery of the body, and are there united by a **circular tube,** which runs parallel with the margin and close to it.

Observe the arrangement of the marginal tentacles; also of the oral tentacles. At the base of the groups of marginal tentacles are minute sense-organs, the **ocelli**; they are characterized by the presence of pigment and are sensitive to light. Note the four swellings on the side of the manubrium; these are the **sexual organs** and are specialized portions of the ectoderm. The sexes are separate; the sexual glands have the same appearance in the two sexes. Around the inner margin of the subumbrella, at the base of the tentacles, is a broad muscular membrane extending inward called the **velum**.

Exercise 7. Make a diagrammatic sketch of the medusa and label all of its parts.

The medusa is a more highly specialized form than the polyp, although they are homologous forms and are essentially alike in structure. The different vegetative functions are carried on in the medusa as they are in the hydranth. The medusa being a free-swimming animal, however, its muscular and nervous systems are much more highly developed than the same systems are in the hydranth. In the latter the only muscles present are delicate fibers, elongated projections of the inner ends of ectodermal cells, which cause movement in the tentacles and the body of the hydranth, while the nervous system is represented only by scattered ganglion cells, which lie among the ectoderm cells. In the medusa the **velum** is the principal organ of locomotion. It contains bands of ectodermal **muscle fibers**, by the contraction of which the motion of the umbrella is produced which propels the animal through the water. The **nervous system** consists of a double **nerve ring** of ganglion cells and fibers running around the margin of the disc, from which delicate fibers run to the velum and to the sense-organs.

HYDROZOA

A CAMPANULARIAN HYDROMEDUSAN (*Obelia* or *Campanularia*)

These are very common marine animals which live in the shallow water along our coast. In common with other members of the group they exhibit the phenomenon of **alternation of generations.** Two generations of individuals, a sexual and an asexual, alternate with each other. The latter is called the **hydroid generation** ; the animal in this stage is sessile and colonial and produces by budding, *i.e.*, by asexual methods, the sexual generation. This latter is called the **medusoid generation** ; in it the animal either remains attached to the hydroid colony (Campanularia) and is then called a **sporosac,** or separates itself (Obelia) and becomes a free-swimming jelly-fish, which is called a **medusa** ; in either case the medusoid produces by sexual methods embryos which attach themselves to fixed objects and develop into the hydroid generation.

The hydroid stage. While in this stage these animals form branching colonies, which are attached to seaweed, rocks, and other objects. Place a small portion of a colony in a watch-glass in water or alcohol, and study it under the microscope. Observe the differences in size between the different polyps, as well as their position on the stem. Determine the method of branching. Has the colony a main axial stem? If not, which is the oldest polyp? Notice the ringed constrictions in various parts of the stem, especially near the polyp; they give the stem strength and flexibility.

The stem of the colony together with the branches is called the **hydrocaulus** ; its root-like projections by which it is attached at its base are the **hydrorhiza.** Observe that there are two distinctly different kinds of polyps instead of but one, as in Pennaria or Bougainvillea: (*a*) the **feeding polyp** or **hydranth,** which is the more numerous and bears tentacles, and (*b*) the **reproductive polyp** or **blastostyle,** which is a modified hydranth and is much the larger and the less numerous and bears no tentacles. Notice that the **perisarc,** the transparent cuticular covering of the stem, does not end at the base of the polyp, as is the case in the tubularian hydroid, but is continued over the polyp, enclosing it as in a cup. It is thus a protective device and is called in the case of the hydranth the **hydrotheca,** and in the case of the blastostyle the **gonotheca.** The feeding polyp withdraws within its hydrotheca for protection when alarmed. The reproductive polyp never emerges from its gonotheca, which is a closed structure, but the medusoids or their sexual products escape into the surrounding water through an opening which finally appears in the gonotheca's free end.

Exercise 1. Draw a diagram representing the method of branching of the colony and the arrangement of the polyps.

Mount a portion of a branch with several hydranths in water or dilute glycerine. Study an expanded hydranth. We note the radial type of structure and the tubular body, the internal cavity of which opens to the outside through the terminal **mouth**; also that the stem has a cavity which is continuous with that of the hydranth. The internal cavity of the hydranth and of the stem is called the **gastro-vascular space,** and is the common digestive and circulatory cavity of the animal. The proboscislike portion of the hydranth between the base of the tentacles and the mouth is called the **hypostome.** Count the **tentacles.** Note the absence of medusoid buds on the hydranth.

Exercise 2. Make a semidiagrammatic sketch of the expanded hydranth on a large scale and label all of its parts.

Exercise 3. Make a sketch of a contracted hydranth.

Find a hydranth bud and study its structure.

Exercise 4. Make a semidiagrammatic sketch of it.

Study the finer structure of an expanded hydranth. First study the structure of a tentacle. It is made up of an axis consisting of a single row of large **entoderm cells**, around which is a layer of small **ectoderm cells**. Between these two cell layers is the delicate non-cellular **supporting layer**. Find the highly refractive **nematocysts** at the end of the tentacle. These are the stinging organs with which the animal kills its prey. Each one consists of a spiral, thread-like tube, with several barbs at its base, which lies coiled within the cavity of a cell called the **cnidoblast**. The cavity is filled with a poisonous fluid; its walls form an ovoid sac, of which the tube is the very much elongated and invaginated outer end. A minute spine projects beyond the free surface of the cnidoblast into the water. When the surface of the ectoderm is irritated the tube is evaginated and violently shot out, and the poisonous fluid contained in the cavity is injected into any animal that may be struck. Look for nematocysts which have discharged their spiral threads.

Exercise 5. Draw the distal portion of a tentacle showing its cellular structure; show the nematocysts at the end, including several which have been discharged.

Study the finer structure of the wall of the hydranth. It is made up of an outer **ectoderm** and a much thicker inner **entoderm**, each composed of a single layer of cells; the inner ends of the entoderm cells are amoeboid and often flagellate, the function of the flagella being to maintain in circulation the fluids in the gastro-vascular space; between these two layers is the thin non-cellular **supporting layer**. The **hydrotheca** encloses all, but it is not in contact with the ectoderm. Study the structure of the stem; it has essentially the same structure as the hydranth;

note the outer cuticular covering, the **perisarc**.   Note the action
of the flagella in a live specimen.

Exercise 6. Make a drawing showing the cellular structure of
the wall of the hydranth and of the stem.

Study a **blastostyle**.   We note that it is a cylindrical object
enclosed within its transparent **gonotheca**.   Budding out on the
sides are the young disc-like **medusæ,** those towards the free end
being the largest and the oldest.   The blastostyle has no
tentacles and no mouth.   It has an internal cavity which is
a part of the gastro-vascular space of the colony, and within
which the nutritive fluids circulate.

Exercise 7. Make a drawing of a blastostyle.

Special respiratory, excretory, digestive, and circulatory organs
are not present in the hydroid.   **Respiration** and **excretion** are
carried on through the surface of the body-wall.   **Digestion,
circulation,** and **absorption** go on within the gastro-vascular space.
The colony lives upon small swimming animals, which the
feeding polyps kill or stun with their nematocysts, and then
swallow into the gastro-vascular space.   Digestion goes on
within the feeding polyps; the products of digestion mingle
with the water present in the gastro-vascular space and circulate throughout the colony.   The entire colony is thus nourished, and if conditions are favorable it will grow rapidly and
produce a large number of medusæ.   The polyps are frequently
destroyed by frost or the beating of waves or by fishes, but new
ones quickly grow in their places.

The medusoid stage.   The medusoids of campanularian hydro-
medusans are either sessile sporosacs or free-swimming medusæ.
The medusæ are minute disc-shaped jelly-fishes, about one-
eighth of an inch in diameter, which may be found swimming in
the surface waters of the ocean.   Place several in a watch-glass
of sea water, or, if they are preserved specimens, in alcohol.

If they are alive, observe the swimming motions. Note the radiate type of structure. The body resembles an umbrella in shape, having a convex and a concave side, and is bordered by a fringe of **tentacles**. The former is called the **exumbrella** or **aboral side**, the latter, the **subumbrella** or **oral side**. In the center of the latter side is the proboscis-like projection called the **manubrium**, at the distal end of which is the **mouth**. This opens into the **gastro-vascular space**, which comprises the space within the manubrium and also a system of canals in the disc-like body. These canals consist of four or more **radial tubes**, which extend from the base of the manubrium to the periphery of the disc, and are there united by a **circular tube** which runs parallel with the margin of the disc and close to it.

Count the marginal tentacles. At the base of certain of the tentacles are minute sense-organs, called **lithocysts**, which are probably organs of equilibrium. Find them.

Near the middle of each radial tube notice a prominent swelling on the subumbrella. These are the **sexual glands** and are specialized portions of the ectoderm. The sexes are separate in medusæ; the sexual glands have the same appearance in the two sexes.

Around the inner margin of the subumbrella, at the base of the tentacles, is a muscular membrane extending towards the manubrium called the **velum**. In campanularian medusæ it is often very narrow and not easily seen ; in tubularian medusæ it is broad and very noticeable.

Exercise 8. Make a diagrammatic sketch of a medusa and label all of its parts.

The medusa is a more highly specialized form than the polyp, although they are homologous forms and are essentially alike in structure. The manubrium of the medusa and the hypostome of the polyp do not differ essentially from each other; the tentacles are also homologous structures. The exumbrella of the medusa corresponds, consequently, to the base of the polyp, and just as the latter is attached to the stem at its base, so the medusa is attached to the blastostyle by its exumbrella. The digestive, excretory, respiratory, and circulatory functions are carried on in the medusa as they are in the hydranth. The medusa being a free-swimming animal, however, its muscular and nervous systems are much more highly developed than are the same systems in the hydranth.

In the latter form the only muscles present are delicate fibers, elongated projections of the inner ends of ectodermal cells, which cause movement in the tentacles and the body of the hydranth, while the nervous system is represented only by scattered ganglionic cells, which are also of ectodermal origin. In the medusa the **velum** is the principal organ of locomotion. It contains bands of ectodermal **muscle fibers**, by the contraction of which the motion of the umbrella is produced which propels the animal through the water. The **nervous system** consists of a double **nerve ring** which runs around the margin of the disc and from which delicate fibers pass to the velum and the sense-organs.

HYDROZOA

A TRACHOMEDUSA (*Gonionemus*)

This animal is a better form to study, on account of its larger size, than the minute tubularian or campanularian medusæ. It is a very common medusa at Woods Hole, but its range of distribution is very limited although it has also been found in Long Island Sound.

Place the medusa in a small dish of water, which should be set upon a dark background. The water should be deep enough to permit the jelly-fish to be readily turned over. If it is alive, study the pulsations of the bell, by means of which it swims. Note the inverted position of the animal when at rest. With a simple lens or a compound microscope study its form and color. Note the radiate type of structure. Unlike the bilaterally symmetrical animals, the medusa has no dorsal, ventral, anterior, or posterior side.

The outer, convex surface of the bell-shaped body is called the **exumbrella** or the aboral side, and the concave underside is called the **subumbrella** or the oral side. From the center of the latter extends a large, dark-brown projection called the **manubrium**, at the distal end of which is the **mouth**, surrounded by four re-curved **lips**. At the base of the manubrium is the **stomach**, a four-sided sac from the four corners of which the four straight **radial canals** extend to the periphery of the body, where they are united by the **ring canal**, which runs around the margin of the bell. The radial and ring canals, together with the stomach and the cavity of the manubrium, form the **gastro-vascular space**, the entodermal lining of which is colored brown.

Directly beneath the four radial canals and projecting slightly into the subumbrella space are the four **reproductive organs,** which are also brown in color and present a corrugated appearance. The sexes are separate, but the animals are not dimorphic.

Observe the number and arrangement of the **tentacles,** of which an adult medusa possesses from sixty to eighty. Note the spiral arrangement of the **nettle cells** on each tentacle, and also the **adhesive pad** near its outer end. It is by means of the nettle cells in these pads that the animal anchors itself to seaweeds and other objects when at rest. Note the exact point above the margin of the bell where the tentacles are inserted. In the basal portion of each tentacle is a conspicuous pigmented body; this is a hollow bulb which is connected with the ring canal. Between the tentacles are the **lithocysts** — minute projections from the margin of the bell which are probably equilibrial in function.

Observe the **velum** — the membrane which extends around the inner margin of the bell towards the manubrium. It is the principal organ of locomotion and contains bands of ectodermal muscle fibers by the contractions of which the motion of the bell is produced which propels the animal through the water. Similar bands of muscle fibers are also present in both the subumbrella and the exumbrella.

Exercise 1. Draw a semidiagrammatic view of the exumbrella on a scale of from 5 to 10, showing the tentacles extended and all the organs which have been observed.

Exercise 2. Draw an oblique side view of the animal on the same scale, showing the velum, the manubrium, and all the other organs observed.

Exercise 3. Draw a semidiagrammatic view of the subumbrella on a scale of 5 to 10, showing the velum, the manubrium, and the other organs observed.

The hydroid generation of Gonionemus is a minute solitary polyp which lives attached to the bottom in shallow water; it will not be studied here. The polyp is only about one millimeter in height and has four tentacles which can be extended two millimeters. It is thus very much smaller than the medusa, which has a height of about nine millimeters and a diameter of about twenty. The polyp forms new polyps by budding, but has never been observed forming the medusæ, so that it is not known how these originate.

ANTHOZOA

A SEA ANEMONE (*Metridium*)

This animal, which is the largest sea anemone along the North Atlantic coast, is often plentiful on rocks, shells, and docks in shallow water. Place an expanded individual in a deep dish of water and observe its shape, color, and method of attachment. The upper end of the columnar body is called the **disc**, and in its center is the elongate, slit-like **mouth**, surrounded by the numerous **tentacles**. The lower end of the animal is called the **foot**. It is often expanded and is not permanently attached to the substratum; the animal has some locomotory powers and can slowly move from place to place.

Study the form of the mouth. Note the thickened lips at each angle of the mouth; these form a ciliated groove, called the **siphonoglyph**, through which the genital products reach the outside. In some individuals only one siphonoglyph is present.

Study the surface of the disc and the tentacles. The former is frequently expanded and thrown into folds and lobes. The tentacles are elongated diverticula of the disc and are hollow. They are charged with nettle cells and are the principal organs of defense and offense. They are also useful in feeding; after the nettle cells have stung the small animals which constitute the food of the sea anemone, the tentacles place them in its mouth. The tentacles are not all the same size, those nearer the mouth being the larger and the older.

Note the character of the columnar body. It is pierced by small pores through which long, white, glandular threads, armed with nettle cells and called **acontia**, may be thrust when the animal is irritated.

Exercise 1. Draw the expanded animal, showing the column and
the disc, with the mouth and the tentacles.

Internal anatomy. Cut the animal into halves by a longitudinal
incision passing at right angles to the mouth and from the disc
to the foot. The mouth will be seen to open into a flattened
tube with more or less corrugated walls, called the **gullet**. Note
the formation of the siphonoglyphs. The gullet leads into the
gastro-vascular space, which is the general internal cavity of the
animal.

The most prominent structures in this cavity are the **mesenteries**,
which are longitudinal partitions extending from the outer wall
of the body inward toward its center. These mesenteries will
be seen to occur in pairs; six of these pairs, called the **primary
mesenteries**, join the body-wall with the wall of the gullet. The
pair at each angle of the gullet which enclose the siphonoglyphs
between them are called the **directives**. Between the six pairs of
primary mesenteries are **secondary**, **tertiary**, and **quarternary** pairs.
The gastro-vascular space is thus divided into a large number
of partially separated longitudinal chambers.

Note carefully the structure of the free edges of the mesenteries below the gullet. The thickened corrugated structure
which forms the edge is the **mesenterial filament**; it contains
digestive glands. From the base of the mesentery extend the
acontia. The reproductive organs, the testes and ovaries, are
also located in the mesenteries, lying alongside the mesenterial
filaments.

Note carefully the position of the longitudinal **muscle bands**,
one of which is present on the surface of each mesentery. It is
by means of these muscles that the body is contracted. A **circular muscle** in the disc closes the mouth by its contraction and aids
in drawing in the tentacles.

Exercise 2. Draw a semidiagrammatic view of the cut surface of
the animal, showing these features.

Make a cross section through the gullet and study the arrangement of the mesenteries, the relation of the primary mesenteries to the gullet, and the longitudinal muscles.

Exercise 3. Draw a diagram of the cross section, showing these features.

Make a cross section through the body beneath the gullet.

Exercise 4. Draw a diagram of the cross section, showing the arrangement of the mesenteries.

CHAPTER IX

SPONGIARIA

CALCAREA

A SYCON SPONGE (*Grantia*)

Grantia is a non-colonial sponge which is common along the New England coast. It is a small cylindrical animal, about half an inch in length, and occurs in small groups attached to rocks or other objects below low-water mark.

Place several specimens in a watch-glass of alcohol or water, and study their shape and external characters with the aid of a hand lens. Observe the cylindrical body and at one end of it a small opening surrounded by straight, needle-like spicules; the opposite end is the one by which the animal was attached. The opening is called the **osculum or excurrent opening.** Notice the smaller spicules and the openings of numerous minute pores which cover the sides of the body. Growing out from the base of the larger individuals may often be seen small ones, which will become, in the course of time, independent animals. Note the evident radial symmetry of the animal.

Exercise 1. Make a drawing of an animal on a scale of 5.

Split a dried sponge with a sharp knife into two equal halves and study it under a dissecting microscope. Observe the large **central cavity.** Large numbers of openings will be seen in its wall; they are the mouths of the **radial canals,** which are. projections of the central cavity into the body-wall. Examine carefully the cut edges of the body-wall; observe the radial canals, which are cut longitudinally here. Notice also the shorter and less regular **incurrent canals**, which lie between the radial canals and open to the outside through external **incurrent pores**. There are thus two systems of canals in the body-wall, (*a*) the **radial canals**, which are a part of the central cavity, and (*b*) the **incurrent canals**, which open to the outside. These two systems of canals communicate with each other by means of minute openings, so that water which enters the incurrent canals from the outside through the external incurrent pores passes freely into the radial canals, and thence into the central cavity. From here it passes out through the osculum.

Exercise 2. Make a semidiagrammatic drawing of the inner surface of the body-wall and the cut edge of the animal, showing the features above described.

Isolate the **spicules** of a sponge by boiling a portion of it in a caustic potash solution. Mount some of them in water and examine them under a high power of the microscope. Find the three different kinds of spicules — the **long straight** ones which guard the osculum, the **short straight** ones which guard the external incurrent pores, and the **triradiate** ones which are within the body-wall and give it rigidity and firmness; some of the latter project into the central cavity. Determine whether the spicules are solid or hollow.

Exercise 3. Draw an outline of each sort of spicule on a large scale.

Make thin sections of a sponge by placing it between two pieces of elder-pith or of cork, and shaving off the sections with a sharp razor or scalpel. Obtain in this way cross, longitudinal, and tangential sections. Mount them in dilute glycerine and study them under the microscope.

Study a cross section in which the canals have been cut longitudinally. Observe the radial and the incurrent canals and their relations to one another. Note the arrangement of the spicules which guard the incurrent pores, also of those triradiate spicules which project into the central cavity.

Exercise 4. Make a drawing of several canals showing these features.

Study a tangential section in which the canals appear in cross section and study the arrangement of the triradiate spicules around them.

Exercise 5. Make a drawing illustrating it.

Specialized **reproductive organs** are not present in Grantia. The sexual elements will be found in the form of large spherical bodies buried in the wall of the sponge. Fertilization takes place here, and development begins, and the young embryos escape into the sea water through the canals. For a while the embryo is a free-swimming animal, but it finally fastens itself to a rock and develops into the adult sponge. Besides this sexual reproduction, the sponge also reproduces asexually by budding. Each distinct cluster of individuals probably represents the gemmated progeny of a single individual.

Special respiratory, excretory, digestive, circulatory, nervous, and locomotory organs are wanting in Grantia. **Respiration** and **excretion** are carried on through the entire surface of the body. The animal feeds on minute organisms and particles of organic matter suspended in the water which streams into the canal system through the incurrent pores. The radial canals are lined with peculiar entoderm cells called **collar cells,** each one of which possesses a **flagellum.** The action of the flagella produces the current of water through the canals, from which the collar cells obtain and ingest food particles. Circulation is from cell to cell.

CHAPTER X

PROTOZOA

INFUSORIA

A FREE-SWIMMING CILIATE INFUSORIAN (*Paramecium*)

Paramecium, often called the slipper animalcule, is one of
the commonest of the larger infusorians. It is a minute, single-
celled animal, being just on the limit of vision, and is almost
universally present in standing water which contains decaying
vegetable matter. It is easily obtained by permitting vegetable
matter to stand in water for a week or two. In shape it is an
elongated ellipsoid with a wide, slightly twisted, longitudinal
groove, called the **oral groove,** on one side; the surface which contains the groove may be called the ventral surface, and the opposite surface, the dorsal. The animal is colorless and transparent,
except when it contains within its body colored food particles.

Mount a drop of water containing Paramecia and some decaying matter on a slide, using a large, thick cover-glass, and study
the animals under a low power of the microscope. They will be
seen swimming rapidly about, but will gradually collect about
the decaying matter. If they do not become quiet in a few
minutes, it is because there is too much water under the cover-
glass, and some of it should be withdrawn with a piece of
blotting paper. Care should be taken that the water does not
all evaporate.

Observe the unsymmetrical shape of the animal, and the
difference between the anterior and the posterior ends. Notice
the rolling over of the animal as it swims through the water;

the peculiar spiral twist of the body is correlated with this motion, but does not necessarily cause it, as the animal may at times revolve in the direction opposite to that of the twist. It is in consequence of this peculiar revolving motion that the animal is able to maintain a course through the water which is practically straight. The great majority of swiftly moving animals are bilaterally symmetrical, and move in straight lines because of that feature of their structure, but Paramecium, together with most free infusorians, has an unsymmetrical form and would tend to move in circles in consequence, without making progress, if it were not for the revolution of its body on its long axis.

Exercise 1. Draw several simple outlines of the body showing its shape as seen in different positions.

Exercise 2. Draw an outline of an ideal cross section through the middle of the body.

Study the structure of the body, using a high power of the microscope when necessary. Study the action of the hair-like vibratile **cilia** which cover the outer surface of the animal and by means of which it moves. They are usually difficult to see in the live animal because of their very rapid motion, but by varying the light and the focus of the microscope they will be brought into view, and in the dead animal are plainly visible. Determine the direction in which the cilia move. Are they all of the same length? Note the delicate transparent **cuticula** which covers the body; it appears as a highly refractive line.

The body has no internal cavity, and the protoplasm of which it is composed is in two distinct layers, the **ectosarc** and **entosarc.** The former is the thick, firm, transparent outer layer which, with the cuticula, gives permanent shape to the body; it often appears obliquely striated. The entosarc is a semifluid granular mass which forms the remainder of the body. From near the anterior end the **oral groove** runs obliquely along the ventral side of the body to a point back of the middle, getting deeper as it goes.   At its inner end the groove becomes a closed tube, which extends into the entosarc and ends with the **mouth.** Notice the **trichocysts** — slender, radially arranged bodies which fill the ectosarc.   They are organs of defense, which remind one of the nematocysts of the Cnidaria; when the animal is irritated they discharge long, delicate bristles, which project beyond the cilia or may leave the body.

Observe the granular nature of the entosarc, and the spherical **food vacuoles** within it.   These are particles of food, usually composed of vegetable substances surrounded by water, which circulate within the semifluid entosarc.   Watch the entosarc closely, and observe the currents in it.   Determine the direction of the currents and whether the direction is ever changed. The food vacuoles form at the inner end of the oral groove, where the particles of which they are composed have been swept by the cilia of the groove.   Watch the formation of them.

Observe the **pulsating vacuoles.**   These are the **excretory organs** of the animal.   They are globular drops of clear liquid, two in number, which appear near the aboral surface of the body, not far from either end, and break through the ectosarc into the surrounding water.   They do not appear simultaneously, but alternate with each other.   When a vacuole has disappeared, radiating canals of clear fluid gradually form about the spot where it was located, bringing the fluid which is to form the next vacuole at that end.   Time the formation of the pulsating vacuoles; how many form in a minute ?

Observe the **macronucleus,** a large, ovoid structure near the center of the body.   At its side are either one or two minute **micronuclei,** according to the species, *P. caudatum* having one and *P. aurelia* two; they may be seen if the animal be killed by adding a 1 per cent. solution of acetic acid to the water.

Exercise 3. Make a large semidiagrammatic drawing of a Paramecium, showing all these details, and label all.

Paramecium has no special vegetative organs except the pulsating vacuoles. Food vacuoles are taken into the entosarc through the mouth. Here they circulate for some time, while the water forming the vacuole is absorbed and the food particles that it contains are digested. The indigestible matters are collected at a spot just back of the mouth and are there ejected from the body through a temporary opening in the ectosarc, which forms for that purpose; the water of the food vacuole is collected in the pulsating vacuoles and ejected. **Respiration** is carried on through the external surface of the body. The organs of **locomotion** are the cilia, which are distributed evenly over the surface of the body; they are hair-like projections of the ectosarc through pores in the cuticula. **Sensation** is exercised by the entire surface of the body.

Reproduction is asexual, by division. A transverse constriction appears in the surface of the middle of the animal's body and deepens until it is divided in two. Each half becomes an independent animal and grows to full size. Look among a large number of animals for one which is dividing.

A process which is universal among infusorians is **conjugation**. Two individuals place the ventral surfaces of their anterior ends together. In this position their bodies fuse together and an interchange of micronuclear matter takes place between them. The two individuals then separate.

Conjugation was formerly supposed to be a process by which weak and infertile animals renewed their strength and vitality. It is now supposed to be rather a preparation for unfavorable life conditions. The change in the structure of the micronucleus leads to a change in the essential characters of the animals, and thus gives them additional powers of environmental adaptation and a better chance to survive unfavorable conditions.

Exercise 4. Look for dividing and also for conjugating individuals. Observe them carefully and draw them.

INFUSORIA

A SESSILE CILIATE INFUSORIAN (*Vorticella*)

This infusorian differs from Paramecium in being a sessile animal, and in that the cilia are not equally distributed over all parts of the body but are confined to certain parts of it. Vorticella and its allies are often called bell animalcules. The animal consists of a **bell-shaped body** at the end of a **long stalk** which is permanently attached to some object in the water. Around the upper and wider margin of the body is a row of large cilia. A deep **oral groove**, which is also bordered by cilia, extends from the margin towards the center of the animal and bears the **mouth** at its inner end.

A number of genera of bell animalcules are found in both fresh and salt water. Vorticella is non-colonial and possesses a contractile spiral stalk ; Carchesium and Zoöthamnium are colonial and differ from each other in that in the former each individual animal contracts independently, while in the latter the entire colony always contracts as a unit; in both, the colonies are large and easily visible, appearing often like white mould on the object of attachment; Epistylis is colonial with a non-contractile stalk.

Mount a drop of water on a slide, together with some vegetable or other substance to which Vorticella is attached, and study it under the microscope. (Any other bell animalcule will do equally well.) Observe the shape of the animal; tap on the slide with a pencil and cause it to contract; note the marginal cilia and the current they set up in the water; find the oral groove and note that the current in the water tends to sweep small objects into it.

Notice the partial radial symmetry of the animal; this body-form is due to its sessile habit of life. Paramecium, which is a rapidly moving animal, is not radially symmetrical. Can you explain why a sessile organism tends to be radial?

Exercise 1. Draw a careful outline of the expanded animal on a large scale, and another of the contracted animal, and label the parts above mentioned.

Study the structure of the body. It consists of a single cell, as does Paramecium, and is composed of two protoplasmic layers, **the ectosarc**, which is the firm external layer, and the **entosarc**, the more fluid protoplasm of which the inner portion of the animal is composed. Covering its outer surface is the **cuticula**, which, with the ectosarc, gives the animal its permanent shape. The stalk is a continuation of the ectosarc and of the cuticula. Its inner portion alone, *i.e.*, the axis, is contractile; its cuticula simply accommodates itself by assuming a spiral shape. Note the longitudinal striations in the ectosarc at the base of the bell.

Observe the granular nature of the entosarc and the spherical **food vacuoles** within it; note the circulation of the latter in the granular protoplasm. Each food vacuole is composed of particles of organic matter in a minute globule of water, which collect in the oral groove and are then driven into the mouth. Watch the formation of them; this is done easily by placing grains of indigo or carmine in the water.

Vorticella has a single **pulsating vacuole**, which is in the upper part of the body. It is the organ of **excretion** of the animal and consists of a globule of clear liquid which collects near the surface of the body and is then discharged through the ectosarc into the water. As in Paramecium, the water which is ingested as a part of the food vacuoles is discharged through the pulsating vacuole together with renal products. Time the formation of the pulsating vacuoles; how many form a minute?

Observe the **macronucleus**; it is a narrow elongated structure and is easily seen; near it is the small spherical **micronucleus**.

Exercise 2. Make a large semidiagrammatic drawing of a Vorticella, showing these details, and label all.

Vorticella has no special vegetative organs except the pulsating vacuole. The food particles which are ingested into the entosarc are there digested, and waste matters are egested through a temporary anus in the upper portion of the body. **Respiration** is carried on through the external surface of the body. Organs of **locomotion** are present in the cilia, by which the animal can swim about if it is broken from its stalk. The axial fiber in the stalk is a delicate striated muscle fiber. **Sensation** is exercised through the external surface.

Vorticella **reproduces** asexually, by a longitudinal division. The process begins at the upper end of the body and proceeds to the base, so that finally there are two individuals upon a single stalk. One of these now separates itself from the stalk, assumes a cylindrical form, and, having developed a band of temporary cilia near one end, swims away to find a place for itself. It soon attaches itself, loses the temporary cilia, and develops a stalk.

In the case of the colonial Vorticellidæ both of the individuals produced by the process of division remain on the stalk. In Zoöthamnium the colony is dimorphic; it contains nutritive individuals which are similar to Vorticella, and reproductive individuals which are large and globose and do not feed. The latter separate themselves from the parents and swim off and found new colonies. This dimorphism and division of labor remind one of the Hydromedusæ. In Vorticella, as in Paramecium, reproduction is largely a matter of sufficient nutrition, well-nourished animals reproducing faster than poorly nourished ones. Conjugation also occurs; it is brought on by the same conditions as in Paramecium and is highly important to the well-being of the race. The process is, however, somewhat different from conjugation in Paramecium. An individual divides into from two to eight parts. These free themselves from the stalk, acquire each a basal band of cilia, and swim about in the water until they come in contact with individuals of the ordinary kind, with which they fuse. A permanent conjugation is then effected instead of a temporary one as in Paramecium.

Conjugation, it will be noticed, while it is not a sexual process, is closely allied to such a process, and it is probably through it that sexuality arose in the organic world. In Paramecium and Vorticella we have two important steps in the development of sexuality. In the former animal the conjugating individuals are of the same size, or isogamous, and the fusion of the two individuals is temporary, while in the latter they are of different sizes, or heterogamous, and the fusion is permanent. As a result of this differentiation in Vorticella one of the conjugating individuals is a large, passive form, while the other is a small, active, motile form, which finds and fuses with the passive form. A distinct foreshadowing of the two sexes which characterize the Metazoa is thus present.

Exercise 3. Look among a large number of Vorticellas for conjugating and for dividing individuals. Observe them carefully and draw outlines of those observed.

MASTIGOPHORA

A FLAGELLATE (*Euglena*)

This single-celled organism, which combines the characters of animals and plants, is often so plentiful in pools and ditches that it makes the water green. It is a minute elongated protozoan, one end of which is pointed and the other blunt; in the latter end is a deep **depression**, from the bottom of which springs a long, thread-like, vibratile **flagellum**. The body is covered by a very delicate **cuticula**; an oral groove and a mouth are not present. The animal is colored green by the presence of **chlorophyll** in its body.

Mount a drop of water containing Euglena on a slide and study it under the microscope. Observe its shape and color; also its swimming motions and the motions of the flagellum. The latter organ will be seen to be at the anterior end of the body; it is always in advance as the animal swims. In some flagellates the flagellum is at the posterior end. Whether the flagellum in any species is at the anterior or the posterior end of the body depends upon the direction the vibratile motion of the flagellum takes. If the motion begins at the base of the flagellum and proceeds towards its tip, the animal's body will be driven ahead with the flagellum at the rear, while if the motion begins at the tip of the flagellum, the body will be drawn after it. Note the extreme plasticity of the body. It can assume a variety of shapes, and will often be seen swimming by the alternate contraction and expansion of the body, like a worm.

Exercise 1. Draw a number of simple outlines of the body showing its shape at different times.

Study the structure of the body. The protoplasm composing it is clear, its surface often showing delicate striations. Note the **cuticula**. In the middle of the body is a spherical **nucleus**. At the anterior end near the depression is a clear space called the **reservoir**; find it. It receives the discharges of the **pulsating vacuole**. This vacuole is a minute globule of clear liquid, which represents the excretory wastes of the animal; it collects and discharges into the reservoir periodically, which thus acts as a urinary bladder and in turn opens into the anterior depression. Near the reservoir is a red **pigment spot,** which is sensitive to light; it is the most primitive form of an **eye.**

Exercise 2. Draw Euglena on a large scale with the above-mentioned organs.

In its life processes Euglena partakes of the nature of both a plant and an animal. Through the agency of the chlorophyll a starch-like carbohydrate called **paramylum** is manufactured, which constitutes a large part of the food of the organism. The process goes on only during the daytime and is a characteristic plant process. But Euglena also ingests solid food after the manner of animals. Food particles are taken into the depression at the anterior end and thence sink into the soft protoplasmic body. **Excretion** is effected through the pulsating vacuole; **respiration,** through the body-surface.

From time to time Euglena encysts itself. It loses its flagellum, draws itself together into a spherical form, and secretes a **cyst** of cellulose. After a while it either throws off the cyst and assumes its former shape or reproduces by dividing into from two to eight small Euglenas. **Reproduction** thus takes place during the period of encystment; also at times free individuals reproduce by longitudinal division.

Exercise 3. Among a large number of individuals look for dividing and also for encysted ones. Make large drawings of several.

SARCODINA

A NAKED RHIZOPOD. AMOEBA

The amoeba is a jelly-like, single-celled animal which may be found in stagnant water attached to submerged objects, or in bottom sediment; it is also often found in moist, damp places which are not under water. The animals are very variable in size, the largest being within the range of the unaided vision, the smallest species requiring high powers of the microscope to detect.

Mount on a slide a drop of water with sediment or scrapings from a submerged leaf or stick containing amoebas, and find one. Observe its shape and granular appearance. From time to time the shape of the body changes by the thrusting out of projections called **pseudopodia.** Observe the formation of pseudopodia.

Exercise 1. Draw several outlines of the animal, showing its shape at different times.

Observe the structure of the body. The protoplasm forming it will be seen to be divisible into two layers, the **ectosarc** and the **entosarc**; the former is the clear, transparent layer which forms the periphery of the body; the latter is the granular, translucent mass which forms the remainder of it. The ectosarc is of firmer consistency than the entosarc and secretes a delicate cuticula on its outer surface. When a pseudopodium begins to form, it consists at first of ectosarc alone, but entosarc finally enters it as it grows larger. The entire body will often flow into a single pseudopodium, in which case

the animal flows in that direction. When this happens the
ectosarc of the hinder portion of the body will be seen to
wrinkle as the entosarc flows away from it.

Observe the granular nature of the entosarc and the flowing
of the granules as they move about with the motion of the
protoplasm. Observe the **food vacuoles** in the entosarc; they are
particles of food surrounded by water. Observe **the pulsating
vacuole**, the organ of excretion. It will be seen to be a large
globule of clear liquid which forms near the periphery and then
discharges into the surrounding water. Time its pulsations;
how many form a minute? Add a 1 per cent. solution of acetic
acid to the water and find the **nucleus**.

Exercise 2. Make a large semidiagrammatic drawing of an amoeba,
showing the features above mentioned, and label all.

Amoeba has no special vegetative organs except the pulsating vacuole. Solid food consisting of plants and animals and
particles of organic matter is ingested in the form of food
vacuoles. These move about in the entosarc with the movements of the animal's body and the nutritive matters are digested
and absorbed. Waste matters are then egested by being thrust
out of a temporary opening in the ectosarc into the water.
Respiration is carried on through the surface of the body. One
reason for the active throwing out of pseudopodia is the necessity of increasing the relative area of the surface of the body
for respiratory purposes.

Reproduction in Amoëba is carried on by division. The
nucleus first divides; the animal then elongates, and a transverse constriction appears in its middle, which is finally carried
through the body. Two animals are thus formed, each of
which contains half of the nucleus. As in other protozoans,
reproduction in Amoeba is largely dependent upon nutrition.
If the nutritive conditions surrounding them are unfavorable
the animals gradually lose their vitality and reproductive powers

and in the course of time will die. Conjugation also occurs, as in other Protozoa. The processes of division and conjugation apparently do not take place frequently, as they have not been often observed.

About a dozen species of the genus Amoeba are known. The commonest are probably *A. proteus*, a large, often active form with long pseudopodia, *A. verrucosa*, a large, sluggish form with very short pseudopodia, *A. limax*, a small form which flows along without definite pseudopodia, and *A. radiosa*, a small star-shaped form with slender, radiating pseudopodia.

APPENDIX

A SYNOPSIS OF THE CLASSIFICATION OF ANIMALS

PHYLUM I. PROTOZOA

Single-celled animals, aquatic and microscopic.

Class 1. *Sarcodina.* Protozoans with more or less protractile pseudopodia.

Order 1. Rhizopoda. Pseudopodia without axial filament and usually very retractile. Ex. Amoeba.

Order 2. Heliozoa. Freshwater Sarcodina with silicious skeleton and ray-like pseudopodia, each with an axial filament. Ex. Actinospherium.

Order 3. Radiolaria. Marine Sarcodina with silicious skeleton. Ex. Polycystina.

Class 2. *Mastigophora* (Flagellata). Protozoans with one or more vibratile flagella. Ex. Euglena.

Class 3. *Sporozoa.* Protozoans which are internal parasites and have no locomotory organs as adults. Ex. Gregarina.

Class 4. *Infusoria.* Protozoans with cilia or sucking tentacles.

Order 1. Ciliata. Ciliate infusorians. Ex. Paramecium.

Order 2. Suctoria. Infusorians with sucking tentacles. Ex. Acineta.

PHYLUM II. COELENTERATA

Radiate animals with a single, but sometimes branched, internal cavity and no cœlom.

SUBPHYLUM I. **Spongiaria** (Porifera). Sessile, mostly colonial animals without specialized organs or tissues; body-wall pierced by numerous pores or canals and usually stiffened by either calcareous or silicious spicules and either with or without spongin fibers.

Class 1. *Calcarea*. Sponges with calcareous spicules and of simple structure. Ex. Grantia.

Class 2. *Hexactinellida*. Glass sponges with six-rayed silicious spicules. Ex. Euplectella.

Class 3. *Demospongiæ*. Massive sponges with either silicious spicules or spongin fibers or both. Ex. Spongilla.

Subphylum II. **Cnidaria**. Coelenterates provided with nettle cells.

Class 1. *Hydrozoa* (Hydromedusæ). Hydroid polyps and jelly-fish, the former without mesenterial ridges and the latter with a velum.

Order 1. Hydrariæ. Freshwater hydroids of simple structure. Ex. Hydra.

Order 2. Hydrocorallinæ. Coral-like marine hydrozoans. Ex. Millepora.

Order 3. Tubulariæ. Hydroids without hydrotheca; medusæ with gonads on the manubrium. Ex. Pennaria.

Order 4. Campanulariæ. Hydroids with hydrotheca; medusæ with gonads on the subumbrella. Ex. Obelia.

Order 5. Trachomedusæ. Hydroids (when present) minute and of simple structure; medusæ usually large with gonads on the subumbrella. Ex. Gonionemus.

Order 6. Narcomedusæ. Hydroids wanting; medusæ with lobed rim. Ex. Cunina.

Order 7. Siphonophora. Free-swimming colonial hydrozoans. Ex. Physalia.

Class 2. *Scyphozoa* (Scyphomedusæ). Hydroids and jelly-fish, the former with mesenterial ridges and the latter without a velum and often of large size. Ex. Aurelia.

Class 3. *Anthozoa*. Sea anemones and corals; solitary or colonial polypoid cnidarians without medusoid generation.

Order 1. Alcyonaria. Anthozoans with eight mesenterial ridges and eight pinnate tentacles. Ex. Corallium.

Order 2. Zoantharia. Anthozoans with numerous mesenterial ridges and numerous simple tentacles. Ex. Metridium.

Subphylum III. **Ctenophora**. Coelenterates with eight bands of ciliated ridges on outer surface. Ex. Mnemiopsis.

PHYLUM III. VERMES

The lower worms. Animals of primitive structure and without paired locomotory appendages or distinct head.

Subphylum I. **Plathelminthes.** Flatworms; no anus present in most forms and body-cavity filled with a vesicular connective tissue called parenchyma.

Class 1. *Turbellaria.* Mostly free-living flatworms with ciliated outer surface. Ex. Planaria.

Class 2. *Trematodes.* Flukes. Small parasitic flatworms with mostly a branched digestive tract and an anterior mouth. Ex. Fasciola.

Class 3. *Cestodes.* Tapeworms. Elongated, usually segmented parasitic flatworms without digestive tract. Ex. Taenia.

Class 4. *Nemertea.* Nemertean worms. Elongated, mostly free-swimming flatworms with a protrusile proboscis and a ciliated outer surface. Ex. Cerebratulus.

Subphylum II. **Nemathelminthes.** Round or thread worms; mostly parasitic. Ex. Ascaris.

Subphylum III. **Trochelminthes** (Rotifera). Minute, aquatic worms with mouth surrounded by cilia. Ex. Rotifer.

Subphylum IV. **Bryozoa.** Minute, sessile, colonial animals with a ridge bearing ciliated tentacles around the mouth. Ex. Bugula.

Subphylum V. **Brachiopoda.** Sessile, marine, mollusk-like animals with a dorsal and a ventral shell. Ex. Terebratulina.

Subphylum VI. **Phoronidea.** Sessile, marine worms living in tubes and with a tentacular ridge around the mouth. Ex. Phoronis.

Subphylum VII. **Chaetognatha.** Minute, transparent, marine worms with a slender body, two or three pairs of horizontal fins, and paired prehensile bristles around the mouth. Ex. Sagitta.

Subphylum VIII. **Sipunculoidea.** Elongated, marine worms, the anterior portion of which can be invaginated and is usually surrounded by tentacles. Ex. Sipunculus.

PHYLUM IV. ANNELIDA

The higher worms. Elongated, segmented worms which have paired, unsegmented appendages, and a usually distinct head.

Class 1. *Archiannelida.* No parapodia or setæ. Ex. Polygordius.

Class 2. **Chaetopoda.** With setæ, segmentally arranged.

Order 1. Polychaeta. Mostly marine chaetopods with parapodia on which are numerous setæ. Ex. Nereis.

Order 2. Oligochaeta. Earthworms. Mostly freshwater or land chaetopods without parapodia and with few setæ. Ex. Lumbricus.

Class 3. **Hirudinea.** Leeches. Annelids with a sucker at each end and no appendages or setæ. Ex. Hirudo.

Class 4. **Myzostomida.** Disk-shaped parasites of echinoderms with five pairs of parapodia. Ex. Myzostoma.

PHYLUM V. ARTHROPODA

Externally segmented animals with segmented appendages.

Class 1. **Crustacea.** Aquatic, gill-bearing arthropods; two pairs of antennæ present.

Division 1. *Entomostraca.* Small, simply constructed crustaceans with a variable number of body-segments and without abdominal appendages.

Order 1. Phyllopoda. Entomostracans with flat, leaf-like append-ages.

Suborder 1. Branchiopoda. Elongated phyllopods with segmented body. Ex. Branchipus.

Suborder 2. Cladocera. Laterally compressed phyllopods, the body of which is not distinctly segmented and is enclosed in a bivalve shell; second pair of antennæ are swimming organs and project from the shell. Ex. Daphnia.

Order 2. Copepoda. Elongated entomostracans with distinctly segmented body and without gills; the female often carries one or two egg-sacs. Ex. Cyclops.

Order 3. Ostracoda. Minute, laterally compressed entomostracans with entire body enclosed in a bivalve shell. Ex. Cypris.

Order 4. Cirripedia. Sessile, hermaphroditic entomostracans with body enclosed in a calcareous shell; barnacles. Ex. Lepas.

Division 2. *Malacostraca.* Crustaceans with a constant number (20) of body-segments and nineteen pairs of appendages; abdominal appendages present.

Subdivision 1. Phyllocarida. Primitive malacostracans with cara-pace and with leaf-like thoracic feet. Ex. Nebalia.

Subdivision 2. Arthrostraca. Malacostracans with usually seven free thoracic body-segments, and with sessile eyes.

Order 1. Amphipoda. Laterally compressed arthrostracans with gills on thorax. Ex. Gammarus.

Order 2. Isopoda. Dorso-ventrally depressed arthrostracans with gills on the abdomen. Ex. Oniscus.

Subdivision 3. Thoracostraca. Malacostracans with carapace covering the head and all or some of the thorax, and with stalked eyes.

Order 1. Schizopoda. Small thoracostracans with carapace covering entire thorax, and with one pair of maxillipeds. Ex. Mysis.

Order 2. Stomatopoda. Thoracostracans with three free thoracic body-segments and large abdomen. Ex. Squilla.

Order 3. Cumacea. Small thoracostracans with reduced carapace. Ex. Diastylis.

Order 4. Decapoda. Large thoracostracans with carapace covering entire thorax, and with three pairs of maxillipeds.

Suborder 1. Macrura. Elongated decapods with large abdomen. Ex. Homarus.

Suborder 2. Brachyura. Broad decapods with reduced abdomen. Ex. Cancer.

Class 2. **Arachnoidea.** Arthropods lacking antennæ and with body usually consisting of cephalothorax and abdomen.

Division 1. *Xiphosura.* Large marine arachnoideans with a long, spike-like telson. Ex. Limulus.

Division 2. *Arachnida.* Usually air-breathing arachnoideans with six pairs of appendages.

Order 1. Scorpionida. Large arachnids with a long segmented abdomen ending in a poisonous sting. Ex. Scorpio.

Order 2. Palpigradi. Minute arachnids with a long, segmented caudal filament. Ex. Koenenia.

Order 3. Pedipalpi. Arachnids with a constriction between the cephalothorax and the segmented abdomen. Ex. Thelyphonus.

Order 4. Solifugæ. Arachnids with a constriction between the head and thorax. Ex. Galeodes.

Order 5. Pseudoscorpionida. Arachnids without a constriction between cephalothorax and abdomen; pedipalps chelate and very long. Ex. Chelifer.

Order 6. Phalangida. Arachnids with extremely long, slender legs and a segmented abdomen. Ex. Phalangium.

Order 7. Araneæ. Spiders. Arachnids with a constriction between the cephalothorax and the unsegmented abdomen. Ex. Agelena.

Order 8. Acarina. Mites. Arachnids with body not divided into cephalothorax and abdomen, and unsegmented. Ex. Hydrachna.

Order 9. Linguatulida. Parasitic arachnids with ringed, vermiform body. Ex. Pentastomum.

Order 10. Tardigradi. Minute, aquatic arachnids. Ex. Macrobiotus.

Order 11. Pycnogonida. Sea spiders. Marine arachnids with very long legs. Ex. Pallene.

Class 3. **Tracheata.** Air-breathing arthropods with one pair of antennæ.

Division 1. *Onychophora.* Worm-like tracheates with indistinctly segmented body and appendages. Ex. Peripatus.

Division 2. *Myriapoda.* Worm-like tracheates with distinctly segmented body and appendages.

Order 1. Progoneata. Body mostly cylindrical and with two pairs of legs to a segment. Ex. Julus.

Order 2. Chilopoda. Centipeds. Flattened myriapods with one pair of legs to a segment. Ex. Lithobius.

Division 3. *Insecta.* Insects. Tracheates with body divided into head, thorax, and abdomen; with three pairs of legs and usually two pairs of wings.

Order 1. Thysanura. Minute, wingless insects without metamorphosis. Ex. Lepisma.

Order 2. Pseudoneuroptera. Insects with two pairs of net-veined wings, biting mouth-parts, and incomplete metamorphosis. Ex. Dragon fly.

Order 3. Orthoptera. Insects with two pairs of wings (the first pair being usually parchment-like), biting mouth-parts, and incomplete metamorphosis. Ex. Grasshopper.

Order 4. Neuroptera. Insects with two pairs of net-veined wings, biting mouth-parts, and complete metamorphosis. Ex. Ant lion.

Order 5. Coleoptera. Beetles. Insects with two pairs of wings (of which the first pair are elytra), biting mouth-parts, and complete metamorphosis. Ex. Potato beetle.

Order 6. Hemiptera. Bugs. Insects with two pairs of wings, or wingless, with sucking mouth-parts in form of a jointed proboscis, and incomplete metamorphosis. Ex. Aphis.

Order 7. Lepidoptera. Butterflies and moths. Insects with two pairs of scale-covered wings, sucking mouth-parts in form of a long, unjointed proboscis, and complete metamorphosis. Ex. Bombyx.

Order 8. Diptera. Insects with one pair of wings, sucking mouth-parts, and complete metamorphosis. Ex. House fly.

Order 9. Hymenoptera. Insects with two pairs of wings, biting and licking mouth-parts, and complete metamorphosis. Ex. Bee.

PHYLUM VI. MOLLUSCA

Animals without paired locomotory appendages, and with a soft, unsegmented body, which is usually enclosed in a calcareous shell.

Class 1. *Amphineura*. Symmetrical mollusks without a shell or with one composed of eight pieces in a longitudinal row. Ex. Chiton.

Class 2. *Scaphopoda*. Symmetrical mollusks with a cylindrical shell. Ex. Dentalium.

Class 3. *Gastropoda*. Snails. Mollusks with an asymmetrical, spiral shell and a single mantle cavity.

Order 1. Opisthobranchiata. Marine snails with posterior gills. Ex. Aeolis.

Order 2. Pulmonata. Freshwater and land snails, without gills but with lungs. Ex. Helix.

Order 3. Prosobranchiata. Mostly marine snails with anterior gills. Ex. Fulgur.

Class 4. *Pelecypoda*. Symmetrical mollusks with a bivalve shell and paired mantle cavities. Ex. Unio.

Class 5. *Cephalopoda*. Mollusks with a large head, which bears a number of long arms, and with a single mantle cavity.

Order 1. Tetrabranchiata. Cephalopods with four gills and a large convoluted shell. Ex. Nautilus.

Order 2. Dibranchiata. Cephalopods with two gills and either eight or ten arms; shell, when present, concealed in the mantle. Ex. Loligo.

PHYLUM VII. ECHINODERMATA

Radially symmetrical animals with calcareous plates or spicules in the body-wall.

Class 1. *Crinoidea.* Sea lilies. Echinoderms which are sessile throughout life or only as larvæ. Ex. Comatula.

Class 2. *Asteroidea.* Starfish. Flattened, star-shaped echinoderms with an ambulacral furrow on the under side of each ray. Ex. Asterias.

Class 3. *Ophiuroidea.* Brittle stars. Flattened echinoderms with long, vibratile arms and without ambulacral furrows. Ex. Amphiura.

Class 4. *Echinoidea.* Sea urchins. Spheroidal or flattened echinoderms without arms. Ex. Arbacia.

Class 5. *Holothurioidea.* Sea cucumbers. More or less worm-like echinoderms with oral tentacles. Ex. Synapta.

PHYLUM VIII. CHORDATA

Animals with a dorsal central nervous system, an internal skeletal system, consisting in the simplest cases of the notochord, and paired pharyngeal slits and arches.

SUBPHYLUM I. **Enteropneusta.** Worm-like chordates with a large proboscis in front of the mouth. Ex. Balanoglossus.

SUBPHYLUM II. **Tunicata.** Chordates in which the body is enclosed in a tunic; a large pharyngeal chamber and a ventral heart present.

Class 1. *Larvacea.* Minute, free-swimming tunicates with a long tail. Ex. Appendicularia.

Class 2. *Thaliacea.* Free-swimming, transparent tunicates. Ex. Salpa.

Class 3. *Ascidiacea.* Sessile, saccular tunicates, either simple or colonial. Ex. Molgula.

SUBPHYLUM III. **Leptocardia.** Elongated, fish-like chordates, compressed laterally and attenuated at both ends. Ex. Amphioxus.

SUBPHYLUM IV. **Vertebrata.** Chordates with distinct head, bearing organs of special sense, with red blood, and usually with two pairs of appendages.

Class 1. *Pisces.* Fishes. Aquatic vertebrates which breathe by means of gills, and usually with bony scales and paired fins. Ex. Perca.

Class 2. ***Amphibia***. Amphibians. Vertebrates with gills during a part or all of their life, and usually with lungs ; scales mostly absent. Ex. Rana.

Class 3. ***Reptilia***. Reptiles. Vertebrates with body covered with horny scales or plates and without gills. Ex. Coluber.

Class 4. ***Aves***. Birds. Feathered vertebrates whose anterior appendages are wings. Ex. Gallus.

Class 5. ***Mammalia***. Mammals. Hair-covered vertebrates which suckle their young. Ex. Felis.

GLOSSARY

Abdomen : the most posterior of the three body-divisions in arthropods; wasp, 2; fly, 7; grasshopper, 10; caterpillar, 20; spider, 25; crayfish or lobster, 31; crab, 42; sow-bug, 46; amphipod, 48, 50; larval decapods, 51; copepod, 53; Daphnia, 56.

Aboral : the side of the body opposite the mouth in a radiate animal; starfish, 142; sea urchin, 149; medusa, 167, 173; Gonionemus, 175.

Aciculum : a chitinous supporting rod in the parapodia of annelids, 63.

Acontia : long threads armed with nettle cells in sea anemones, 178.

Adductor muscle : a muscle which draws an organ towards the axis of the body; mussel, 90; oyster, 100; clam, 104.

Air-sacs : tracheal enlargements in insects, 17.

Algæ : very simple green plants, 160.

Alimentary tract : the digestive canal, the organ which ingests, digests, and absorbs the food; see Digestive System in Index.

Alternation of generations : the alternate succession of sexual and asexual generations in hydromedusans, 163, 169.

Alveolus : a pyramidal ossicle which supports one of the five teeth in the dentary apparatus of the sea urchin, 152.

Ambulacral feet : tubular projections with sucker discs at their ends in echinoderms, 143, 149, 157.

Ambulacral groove : the elongated groove on the oral side of the rays of the starfish, 143.

Ambulacral pores : minute openings in the body-wall in the starfish, 143; in the sea urchin, 150.

Ametabolic : larval development without metamorphosis in insects.

Ampulla : a sac-like projection of the ambulacral foot in echinoderms, 146, 153, 157.

Anal feelers : paired posterior projections; centiped, 22; sow-bug, 46.

Analogous : having a similar function.

Antenna : a segmented sensory appendage on the head of arthropods; wasp, 2; beetle, 5; fly, 7; grasshopper, 9; caterpillar, 20; centiped, 22; crayfish or lobster, 29; crab, 43; sow-bug, 46; amphipod, 48; copepod, 53; Daphnia, 56; nauplius, 59.

Anterior : at or towards the front end of the body.

Anus : the posterior opening of the digestive canal.

Aorta : a large artery leading directly from the heart; spider, 26; snail, 116; squid, 129.

Appendage : see Extremity; wasp, 1; grasshopper, 12; centiped, 22; spider, 24; crayfish or lobster, 30; crab, 43; sow-bug, 47; amphipod, 49; Caprella, 50; larval decapods, 51; copepod, 54; Daphnia, 56; nauplius, 59; Nereis, 61, 63: a projection from some part of the body.

Appendix : a short diverticulum of the intestine, 40.

Aristotle's lantern : the dentary apparatus of the sea urchin, 151.

Artery : a blood vessel carrying blood away from the heart to the tissues; crayfish or lobster, 37, 38; Nereis, 64; mussel, 94; clam, 108; snail, 116; squid, 127, 129.

Arthrobranch : a gill attached to the joint between the leg and the body in crustaceans, 35.

Articulate : composed of a series of homologous segments.

Asexual : reproduction by division or budding and not through the agency of the sexes; Bugula, 87; Hydra, 161; hydromedusan, 163, 169; sponge, 183; Paramecium, 187; Vorticella, 190; Euglena, 193; Amoeba, 195.

Auricle : a chamber of the heart which receives the blood from the veins; mussel, 94; oyster, 101; clam, 108; snail, 116.

Avicularium : a structure shaped like a bird's head attached to the zoœcium in Bryozoa, 87.

Axial organ : a glandular organ in the axial sinus in the starfish, 148; in the sea urchin, 154.

Axial sinus : an elongated sac alongside the stone canal in the starfish, 148; in the sea urchin, 154.

Balancers : the homologues of the metathoracic wings in Diptera, 8.

Bilateral symmetry : having the right and left sides alike.

Bivalve : a shell composed of two distinct and equivalent parts or valves; mussel, 89; oyster, 99; clam, 103.

Bivium : the two rays of a starfish or a sea urchin which enclose the madreporic plate between them, 142, 151; the two rays on the upper side of the holothurian's body, 155.

Blastostyle : the reproductive polyp of a campanularian hydroid, 170.

Body-cavity : an internal space in the body in which lie the viscera.

Body-wall : the outer portion of the body, which usually bounds the body-cavity towards the inside.

Brachial : relating to the arms.

Branchial : relating to the gills.

Branchial heart : a lateral heart in the squid, 129.

Branchiate : bearing gills.

Branchiostegite : a paired lateral fold of the body-wall in crustaceans, 30.

Brood-sac : a chamber in which the eggs develop in certain crustaceans; sow-bug, 47; amphipod, 49; Daphnia, 57.

Bud : an outgrowth of an animal which becomes a new individual; Bugula, 87; Hydra, 161; hydromedusan, 166, 172, 177.

Calcareous : formed of carbonate of lime.

Capsulogenous glands : glands of uncertain function, but which may aid in the secretion of the cocoon in the earthworm, 69.

Carapace : the shell covering a portion or all of the cephalothorax in crustaceans; crayfish or lobster, 29; crab, 42; larval decapods, 51.

Cardo : the basal division of the maxilla in insects, 13.

Cellulose : the woody cell-wall of plant cells; Molgula, 135; Euglena, 193.

Cephalothorax : a body-division formed by the fusion of the head and the thorax in arthropods; spider, 24; crayfish or lobster, 28; crab, 42; larval decapods, 51; copepod, 53.

Cercus : a paired projection at the posterior end in certain insects, 10.

Cheliped : the large grasping claw in many crustaceans, 30.

Chitin : a hard and very resistant substance present in the cuticula of arthropods.

Chloragogue cells : glandular cells surrounding the digestive canal of the earthworm, 71.

Chlorophyll : the green coloring matter of plants, 192.

Chromatophores : pigment bodies; squid, 123; Hydra, 160.

Cilia : the numerous vibratory projections on the outer surface of certain animals; planarian, 76; Paramecium, 185; Vorticella, 188 : and of certain organs; Bugula, 86; mussel, 94; clam, 108.

Cirrus : a filamentous, sensory appendage of annelids, 62, 63 : a protrusile, copulatory organ of flatworms, 77.

Clitellum : a thickened glandular region on the earthworm which secretes the cocoon, 68.

Cloaca : a tubular or sac-like space which receives the discharge of various organs; planarian, 77; tapeworm, 83; mussel, 92; clam, 106; snail, 121.

Clypeus : a median sclerite in the face of insects just back of the upper lip, 12.

Cnidoblast : a stinging cell in Cnidaria which contains the nematocyst; Hydra, 161; tubularian, 165; campanularian, 171.

Cocoon: a case containing one or more developing animals, 73.

Cœcum: a sac-like appendage of the digestive tract; grasshopper, 15, starfish, 145.

Cœlom: the body-cavity.

Collar: the ventral edge of the mantle in gastropods, 113; in cephalopods, 124.

Collar cells in sponges, 183.

Colon: a division of the intestine in insects, 15.

Columella: the axis of a spiral snail's shell, 113.

Compound eye: an eye made up of a number of separate elements, or ommatidia, in arthropods; wasp, 2; grasshopper, 9; crayfish or lobster, 29.

Conjugation: the fusion of two protozoans and interchange of nuclear matter; Paramecium, 187; Vorticella, 190; Amoeba, 196.

Connective tissue: a tissue whose principal function is to support and hold in place other tissues and organs.

Coxa: the proximal segment of an insect's or a spider's leg, by which it articulates with the body, 4, 12, 25.

Crop: a dilated portion of the œsophagus; grasshopper, 15; earthworm, 71.

Ctenidium: a respiratory organ in mollusks; mussel, 93; clam, 107.

Cuticula: the outer layer of the integument of most invertebrates; wasp, 1; beetle, 5; fly, 7; grasshopper, 10; spider, 24; crayfish or lobster, 28; Nereis, 61; earthworm, 74; Bugula, 85; mussel, 89; clam, 103; snail, 112; Molgula, 136; tubularian, 166; campanularian, 170; Paramecium, 185; Vorticella, 189; Euglena, 192.

Cyst: a capsule containing an animal usually in a state of suspended animation; tapeworm, 83; Euglena, 193.

Cysticercus: a cyst containing a tapeworm scolex, 84.

Dentary apparatus: the five teeth and their supporting structure in the sea urchin, 151.

Development: the series of changes in the early life of an animal by which it passes from the condition of a fertilized egg to that of the adult.

Dimorphism: the condition in which a species exists in two distinct forms, as, for instance, male and female.

Distal: a position away from the point of attachment — opposed to proximal.

Diverticulum: a sac-like projection of a tubular organ.

Dorsal: on or towards the back.

Dorsal lamina: a ciliated ridge in the mid-dorsal line of the pharynx in the ascidean, 139.

Ductus ejaculatorius : the terminal portion of the male reproductive tract in insects, **17.**

Ectoderm : the outermost layer of cells in the Cnidaria and Spongiaria; Hydra, 160; tubularians, 165; campanularians, 171.

Ectosarc : the outermost layer of non-granular protoplasm in protozoans; Paramecium, 185; Vorticella, 189; Amoeba, 194.

Elytra : the hard wing-covers in beetles, 5.

Embryo : a young animal which is passing through its developmental stages, usually within the egg membraues or in the maternal uterus.

Encyst : the act of an animal in forming a cyst about itself.

Endopodite : the innermost of the two terminal branches of the typical crustacean foot; crayfish or lobster, 32; sow-bug, 47; Daphnia, 56.

Endoskeleton : an internal supporting structure.

Endostyle : a ciliated and glandular groove in the mid-ventral line of the pharynx in ascidians, 137.

Entoderm : the innermost layer of cells in the Cnidaria and Spongiaria; Hydra, 160; tubularian, 165; campanularian, 171; sponges, 183.

Entosarc : the inner granular protoplasm in protozoans; Paramecium, 185; Vorticella, 189; Amoeba, 194.

Epicranium : the sclerite forming the dorsal, median, and lateral walls of the head in insects, 12.

Epigynum : a cuticular plate covering or accompanying the female genital pore in many species of spiders, 26.

Epimeral plates : plates which may extend from the latero-ventral sides of the thorax in amphipods, 48.

Epiphragma : the disc of calcified slime with which a land snail can close the opening of its shell, 112.

Epipodite : a membranous projection of the protopodite; crayfish or lobster, 31; crab, 43.

Exopodite : the outermost of the two terminal branches of the typical crustacean foot; crayfish or lobster, 32; sow-bug, 47; Daphnia, 56.

Extensor muscle : a muscle that extends an organ, 33.

Extremity : a paired lateral or ventral appendage of the body of an animal, used primarily for locomotion, although in many cases having secondarily some other function, 1.

Exumbrella : the aboral side of a medusa, 167, 173, 175.

Femur : the segment of an insect's or a spider's leg between the trochanter and the tibia, 4, 12, 25.

Fertilization: the union of the spermatozoön and the ovum.

Flagellum: a vibratory thread-like projection of certain cells; hydroids, 162, 166; sponges, 183; also of flagellate infusorians, 192.

Flame cell: the terminal cell of an excretory tubule of flatworms, 78.

Flexor muscle: a muscle that bends an organ, 37.

Food vacuole: a globule of water containing food particles; Paramecium, 186; Vorticella, 189; Amoeba, 195.

Front: the anterior median portion of the epicranium, 12.

Funiculus: a mesenteric strand connecting the stomach pouch with the body-wall in bryozoans, 86.

Funnel: the siphon of a cephalopod, 124.

Ganglion: an aggregation of nerve cells; grasshopper, 18; crayfish or lobster, 41; crab, 44; Daphnia, 57; Nereis, 66; earthworm, 74; mussel, 97; clam, 111; snail, 121; squid, 133; Molgula, 138.

Gastrolith: a calcareous body sometimes present in the stomach of crustaceans, 40.

Gastro-vascular space: the central cavity in Cnidaria; Hydra, 160; tubularian, 164; campanularian, 170; Gonionemus, 176; sea anemone, 179.

Gastrula: a stage in the development of the embryo in which two cell layers only are present, the ectoderm and the entoderm.

Gena: the lateral portion of the epicranium in insects, 12.

Genital plate: a sclerite at the posterior end of the abdomen in the male grasshopper, 11.

Giant fibers: three large fibers in the dorsal portion of the nerve cord in the earthworm, 75.

Gill: an organ for the breathing of air contained in the water; crayfish or lobster, 35; crab, 43; sow-bug, 47; amphipod, 49; Caprella, 50; Nereis, 63; mussel, 90; oyster, 100; clam, 104; squid, 127.

Gill-filament: ciliated vertical ridges on the surface of the gills of lamellibranchs; mussel, 94; clam, 108.

Gizzard: a portion of the alimentary tract with thickened muscular walls, 71.

Glochidium: the larval form of Anodonta and Unio, which lives a parasitic life in the skin of fishes, 97.

Gonotheca: the cuticular outer covering of the blastostyle, 170.

Green gland: the kidney of a malacostracan crustacean, 30.

Hæmal: pertaining to the blood system.

Head: the anterior body-division of the higher animals.

Heart: a muscular tube-like or sac-like organ which propels the blood; grasshopper, 14; spider, 26; crayfish or lobster, 38; crab, 44; Daphnia, 57; earthworm, 70; mussel, 94; oyster, 101; clam, 108; snail, 116; squid, 129; Molgula, 137; starfish, 148; sea urchin, 154.

Hemimetabolic: larval development with incomplete metamorphosis in insects.

Hermaphroditic: having the two sexes united in one animal; earthworm, 72; planarian, 77; tapeworm, 81; Bryozoa, 87; snail, 120; Molgula, 137; Hydra, 162.

Hinge ligament: the flexible portion of a bivalve shell which joins the two valves; mussel, 89; oyster, 99; clam, 103.

Holometabolic: insects having a complete metamorphosis.

Homologous: having had a similar origin.

Host: the animal which harbors a parasite, 80.

Hydranth: a feeding polyp in a hydroid colony, 164, 170.

Hydrocaulus: the stem of a hydroid colony, 164, 169.

Hydroid: the sessile, asexual generation of the Hydromedusæ, 163, 169.

Hydrorhiza: the root-like projections of a hydroid colony by which it is attached, 164, 169.

Hydrotheca: the cuticular outer covering of the hydranth in campanularian hydroids, 170.

Hypodermis: the cellular layer which forms the inner portion of the integument of most invertebrates; crayfish or lobster, 36; earthworm, 74.

Hypopharynx: a median projection from the ventral wall of the pharynx in insects — in many insects an important mouth-part, 13.

Hypophysis: a ventral projection of the brain in vertebrates, 138.

Hypostome: the projection of a hydroid's body which bears the mouth; Hydra, 160; campanularian, 170.

Ileum: a division of the intestine in insects, 15.

Imago: a holometabolic insect which has completed its metamorphosis; an adult insect.

Integument: the outer covering of an animal; in most invertebrates it consists of an outer cuticula and an inner hypodermis.

Interfilamentary connections: cross-ridges which join the gill-filaments in lamellibranchs; mussel, 94; clam, 108.

Interlamellar partitions: vertical walls which join the two lamellæ of a lamellibranch's gill; mussel, 93; clam, 107.

Intermediate host: the animal which harbors the larval form of a parasite, 84.

Interray: one of the divisions of the radiate body of echinoderms; starfish, 142; sea urchin, 150; sea cucumber, 155.

Intestine: the division of the digestive tract in which absorption goes on; spider, 27; crayfish or lobster, 37; copepod, 55; planarian, 77; Bugula, 86; mussel, 96; oyster, 101; clam, 110; snail, 118; squid, 131; Molgula, 137; starfish, 145; sea urchin, 152; sea cucumber, 156.

Kidney: an excretory organ; spider, 27; crayfish or lobster, 40; Nereis, 65; earthworm, 73; mussel, 95; clam, 109; snail, 116; squid, 127; Molgula, 138.

Labium: the under lip of insects; fly, 8; grasshopper, 12; beetle, 14; wasp, 14; caterpillar, 20; spider, 25.

Labrum: the upper lip of insects and of some crustaceans; grasshopper, 12; beetle, 14; wasp, 14; caterpillar, 20; crayfish, 30.

Lamella: a leaf-like or plate-like structure, 90, 104.

Larva: a young animal which has left the egg and is leading a free life, but which has not yet completed its development; decapods, 51; entomostracan, 59; tapeworm, 84; mussel, 97: a holometabolic insect between the embryonic and the pupal stages, 20.

Lateral: a position to the right or left of the median line.

Ligula: the anterior portion of the labium in insects; grasshopper, 13; wasp, 14.

Lithocyst: a marginal sense-organ in certain medusæ, 173, 176.

Liver: a digestive gland; crayfish or lobster, 39; crab, 44; Daphnia, 57; mussel, 96; oyster, 101; clam, 107; snail, 119; squid, 131; starfish, 145.

Lophophore: a circular or horseshoe-shaped ridge bearing tentacles in Bryozoa, 86.

Lumen: the cavity within a tubular organ.

Macronucleus: the large nucleus of an infusorian; Paramecium, 186; Vorticella, 190.

Madreporic plate: a porous plate through which fluids enter the ambulacral system; starfish, 142; sea urchin, 151; sea cucumber, 157.

Malpighian tubules: the kidney of insects and certain other arthropods; grasshopper, 16, caterpillar, 21; spider, 27.

Mandible: the anterior pair of mouth-parts in arthropods; grasshopper, 13; beetle, 14; wasp, 14; caterpillar, 20; spider, 24; crayfish or lobster, 30; sow-bug, 47; amphipod, 49; copepod, 54; Daphnia, 56; nauplius, 59.

Mantle: the integumental fold in mollusks which secretes the shell; mussel, 90; oyster, 100; clam, 104; snail, 113; squid, 124 : the body-wall of ascidians beneath the tunic, 136.

Manubrium: the projection of a medusa's body which bears the mouth, 167, 173, 175.

Maxilla: the paired mouth-parts immediately behind the mandibles in arthropods; grasshopper, 13; beetle, 14; wasp, 14; caterpillar, 20; spider, 25; crayfish or lobster, 30; sow-bug, 47; amphipod, 49; copepod, 54; Daphnia, 56.

Maxillipeds: the anterior thoracic appendages which assist in mastication in crustaceans; crayfish or lobster, 30; sow-bug, 47; amphipod, 49; Caprella, 50.

Medusa: a medusoid which becomes a free-swimming jelly-fish, 163, 169, 175.

Medusoid: the sexual generation of a hydromedusan, 163, 169.

Megalopa: a larval stage of the crab, 51.

Mentum: a division of the labium in insects, 13.

Mesentery: a lamella which supports some one of the viscera; Nereis, 64; Bugula, 86; squid, 127; starfish, 145; sea urchin, 152; sea cucumber, 156; sea anemone, 179.

Mesosternum: the ventral surface of the mesothorax in insects; wasp, 3; grasshopper, 10.

Mesothorax: the second thoracic somite in insects; wasp, 2; grasshopper, 10.

Metamere: one of the serial, homologous body-segments, together with its appendages, which form the body of an articulate animal.

Metamorphosis: the quiescent period in the life of a holometabolic insect during which it changes from a larva to an imago.

Metasoma: the primitive segmented trunk of an articulate animal, 62, 68.

Metasternum: the ventral surface of the metathorax in insects; wasp, 3; grasshopper, 10.

Metastomium: the posterior portion of the prosoma of an annelid, 62, 69.

Metathorax: the third thoracic somite in insects; wasp, 2; grasshopper, 9.

Metazoa: the division of the animal kingdom comprising the many-celled animals, 191.

Micronucleus: the smaller of the nuclear bodies in infusorians; Paramecium, 186; Vorticella, 190.

Mother-of-pearl: the inner layer of the shell of mollusks; mussel, 91; clam, 105.

Moult: to shed the cuticula or the outer portion of it.

Mouth-parts : the masticatory appendages on the head of arthropods ; wasp, 2 ; grasshopper, 12 ; beetle, 13 ; wasp, 13 ; caterpillar, 20 ; centiped, 22 ; crayfish or lobster, 30 ; crab, 43.

Mysis stage : a larval form of the lobster, 51.

Nauplius : a larval form of crustaceans, 59.

Nematocyst : the stinging organ in the Cnidaria which is within the cnidoblast ; Hydra, 161 ; tubularian, 165 ; campanularian, 171.

Nephridium : a urinary tubule in annelids ; Nereis, 65 ; earthworm, 73.

Nephrostome : the ciliated opening of a uephridium into the body-cavity ; Nereis, 65 ; earthworm, 73.

Nerve commissure : a nerve connecting the two members of a pair of ganglia ; planarian, 78 ; tapeworm, 82 ; mussel, 97 ; clam, 111 ; snail, 121.

Nerve connective : a nerve connecting two ganglia not of the same pair ; grasshopper, 18 ; spider, 27 ; crayfish or lobster, 41 ; Nereis, 65 ; mussel, 97 ; oyster, 102 ; clam, 111.

Nettle cell : the stinging organ in the Cnidaria ; Hydra, 159 ; Gonionemus, 176 ; sea anemone, 178.

Neuropodium : the ventral division of the parapodium of an annelid, 63.

Nidamental glands : the large glands which secrete the egg-capsules in the squid, 133.

Notopodium : the dorsal division of the parapodium of an annelid, 63.

Nucleus : a spheroidal body in a cell, the center of its activities ; Paramecium, 186 ; Vorticella, 190 ; Euglena, 193 ; Amoeba, 195.

Ocellus : a minute primitive eye ; wasp, 2 ; fly, 7 ; grasshopper, 9 ; caterpillar, 20 ; tubularian medusa, 168.

Ocular plate : the plate at the aboral end of a ray of the sea urchin, 151.

Œsophagus : the gullet, the division of the digestive canal leading from the pharynx to the stomach ; grasshopper, 15 ; caterpillar, 21 ; crayfish or lobster, 39 ; Nereis, 64 ; earthworm, 71 ; Bugula, 86 ; mussel, 96 ; clam, 110 ; snail, 117 ; squid, 130 ; Molgula, 137 ; starfish, 145 ; sea urchin, 151 ; sea cucumber, 156.

Ommatidium : a single element of the compound eye of an arthropod.

Oœcium : a structure in Bryozoa in which the embryo develops, 88.

Oral : the side of the body containing the mouth in a radiate animal ; starfish, 142 ; sea urchin, 149 ; sea cucumber, 155 ; medusa, 167, 173, 175.

Oral groove : a groove leading to the mouth in ciliate infusorians, 184, 188.

Organ of Keber : an organ probably excretory in function in lamellibranchs ; mussel, 92 ; clam, 107.

Osculum: the cloacal opening in sponges, 181.

Ossicles: the calcareous plates in the body-wall of echinoderms.

Otocyst: an organ of hearing; mussel, 98; clam, 111.

Ovarioles: the tubules forming the ovary of an insect, 16.

Ovary: the female sexual gland; grasshopper, 16; spider, 27; crayfish or lobster, 37; crab, 44; copepod, 55; Daphnia, 57; earthworm, 72; planarian, 77; tapeworm, 83; mussel, 97; oyster, 102; clam, 110; squid, 133; starfish, 146; Hydra, 161.

Oviduct: the tube leading from the ovary towards the outside; grasshopper, 16; spider, 27; crayfish or lobster, 39; crab, 44; copepod, 55; Daphnia, 57; earthworm, 72; planarian, 77; snail, 120; squid, 133.

Ovipositor: the organ by means of which certain insects deposit their eggs; fly, 8; grasshopper, 10.

Ovoid gland: the axial organ in the starfish, 148; in the sea urchin, 154.

Ovum: the female sexual cell, the egg.

Pallial line: the line along which the margin of the mantle is attached to the shell in lamellibranchs; mussel, 91; clam, 195.

Pallial sinus: the indentation in the pallial line caused by the insertion of the siphonal retractor muscle, 105.

Palp: a sensory organ near the mouth; wasp, 2; grasshopper, 13; crayfish or lobster, 34; Nereis, 62; mussel, 92; oyster, 101; clam, 106.

Pancreas: a digestive gland in the squid, 130.

Papulæ: the delicate projections of the body-wall in the starfish, 142.

Paragnatha: delicate lamellæ just behind the mandibles in the crayfish or lobster, 30.

Paramylum: a granular substance resembling starch in Euglena, 193.

Parapodium: the appendage of annelids, 63.

Parasite: an animal which attaches itself to another and lives upon its nutritive fluids, 80.

Parenchyma: a vesicular connective tissue which fills the body-cavity of flatworms and leeches; planarians, 79; tapeworm, 81.

Parthenogenesis: reproduction by means of unfertilized eggs, 58.

Pedicellariæ: minute pincer-like organs present on the external surface of starfishes, 142; sea urchins, 150.

Pedipalps: the second pair of appendages in the Arachnida, 24.

Pen: the shell of the squid, 125.

Pericardium: the membrane surrounding the heart; spider, 26; crayfish or lobster, 38; mussel, 92; oyster, 101; clam, 107; snail, 115; Molgula, 137.

Periopods: the thoracic appendages posterior to the maxillipeds in crustaceans, 33.

Periostracum: the outer layer of the molluscan shell; mussel, 91; clam, 105; snail, 115.

Periphery: the outer surface of a body.

Periproct: the region immediately around the anus, 150.

Perisarc: the cuticular outer covering of a hydroid; tubularian, 166; campanularian, 170.

Peristome: a membrane surrounding the mouth in echinoderms; starfish, 143; sea urchin, 149.

Peristomium: the posterior portion of the head in most annelids, consisting of the metastomium and the anterior somites of the metasoma, 62.

Peritoneum: the membrane lining the body-cavity.

Pharynx: the division of the alimentary tract immediately back of the mouth; grasshopper, 15; Nereis, 64; earthworm, 71; Bugula, 86; snail, 117; squid, 131; Molgula, 136.

Plankton: a collective term referring to all small forms of life in the surface waters of the sea or of fresh water.

Pleopod: an abdominal appendage in crustaceans, 32.

Pleurobranch: a gill attached to the body-wall in crustaceans; crayfish or lobster, 35; crab, 43.

Pleurum: the lateral surface of the thorax in insects; wasp, 3; grasshopper, 10.

Podical plates: paired sclerites at the posterior end of the abdomen in certain insects, 11.

Podobranch: a gill attached to the leg in crustaceans, 35.

Polian vesicle: a sac extending from the ring canal in echinoderms, 157.

Polyp: a sessile individual in the Cnidaria; Hydra, 159; tubularian, 163; campanularian, 169; Gonionemus, 177.

Polypide: the soft parts of a bryozoan, 85.

Posterior: at or towards the hinder end of an animal.

Proboscis: a prehensile organ in certain animals, usually a portion of the pharynx; Nereis, 64; planarian, 76: the beak-like mouth-parts of certain insects, 7.

Proglottid: a tapeworm segment, 80.

Prosoma: the primitive head of annelids made up of the prostomium and the metastomium, 62, 68.

Prostate gland: the gland which secretes the fluid in which the spermatozoa are suspended; grasshopper, 17; squid, 132.

Prosternum : the ventral surface of the prothorax in insects; wasp, 3; grasshopper, 10.

Prostomium : the anterior portion of the head of annelids, 62.

Prothorax : the first thoracic segment in insects; wasp, 2; grasshopper, 9.

Protopodite : the basal segment of a crustacean's leg; crayfish or lobster, 32.

Protractor muscle : a muscle which extends the organ to which it is attached; Nereis, 64; mussel, 91.

Proximal : a position towards the point of attachment — opposed to distal.

Pseudopodium : a retractile process in rhizopods, 194.

Pulsating vacuole : a globule of excretory fluid in many protozoans; Paramecium, 186; Vorticella, 189; Euglena, 193; Amoeba, 195.

Pulvillus : an adhesive pad on the foot of insects; fly, 8; grasshopper, 12.

Pupa : the stage in the life of a holometabolic insect when it is undergoing its metamorphosis.

Racemose vesicles : minute diverticula of the ring canal in starfishes, 146.

Radial symmetry : having the parts or organs arranged symmetrically about a common center.

Radial tubes : a portion of the gastro-vascular space in the medusa, 167, 173, 175.

Radula : the band of calcareous teeth in the pharynx of gastropods and cephalopods; snail, 120; squid, 132.

Ray : one of the main divisions of the radiate body of echinoderms, 141, 150, 155.

Receptaculum seminis : a receptacle for sperm in the female animal; grasshopper, 16; spider, 27; crab, 44; snail, 118.

Rectal glands : glandular structures in the rectum of certain insects, 15.

Rectum : the posterior division of the digestive tract; grasshopper, 15; caterpillar, 21; crayfish or lobster, 39; Bugula, 86; mussel, 96; clam, 110; snail, 118; squid, 131; sea cucumber, 156.

Respiratory tree : a branched diverticulum of the rectum in holothurians, 156.

Retractor muscles : muscles which draw in an organ to which they are attached; Nereis, 64; mussel, 91; clam, 105.

Rostrum : a projection of the carapace in crustaceans, 29.

Salivary glands : digestive glands at the anterior end of the digestive tract; grasshopper, 15; snail, 118; squid, 131.

Scaphognathite : the elongated epipodite of the second maxilla in certain crustaceans, 34.

Sclerite: a small plate forming a portion of the cuticula of a segment in insects.

Scolex: the anterior end of a tapeworm, 80.

Scutellum: a small sclerite in the tergum of the thoracic segments in insects.

Segment: one of a number of serial divisions of an animal's body or of an organ.

Septum: a plate forming a division wall between two spaces; Nereis, 63; earthworm, 69; mussel, 92; clam, 106.

Sessile: fixed to one place, without locomotory powers — of an animal; Bugula, 85; oyster, 99; Molgula, 135; tubularian, 163; campanularian, 169; Grantia, 181; Vorticella, 188: not on a stalk or stem — of an organ, 46.

Seta: a bristle; Nereis, 61; earthworm, 67.

Setigerous glands: glands which secrete setæ, 73.

Sexual: reproduction through the agency of the two sexes.

Shell gland: the kidney of entomostracans; copepod, 55; Daphnia, 57.

Siphon: the organ through which water enters or leaves the mantle cavity in mollusks and ascidians; mussel, 92; clam, 106; squid, 124; Molgula, 135.

Siphonoglyph: a ciliated groove in the angle of the gullet in Anthozoa, 178.

Somite: one of the serial, homologous body-segments which form the body of an articulate animal.

Spermatheca: a sac for the storing of sperm in the female animal, a seminal receptacle, 72.

Spermatophore: a capsule or mass of spermatozoa; copepod, 55; snail, 121; squid, 133.

Spermatozoön: the male sexual cell.

Sperm-duct: the vas deferens, 27, 71.

Sperm-sac: a sac for the storing of sperm in the male animal, a seminal vesicle, 71.

Sperm-sphere: a mass of spermatozoa in the earthworm, 73.

Spicule: a minute calcareous or silicious body in sponges and echinoderms.

Spinnerets: the appendages on a spider from which the silk exudes, 25.

Spiracle: an external opening in the tracheal system; wasp, 3; fly, 8; grasshopper, 11; caterpillar, 21; spider, 26.

Sporosac: a sessile medusoid, one which remains attached to the parent hydroid; tubularian, 163; campanularian, 169.

Sternite: the ventral portion of an abdominal segment in insects; wasp, 3; grasshopper, 10.

Sternum : the ventral surface of the thorax in arthropods; wasp, 3 ; grasshopper, 10 ; spider, 25.

Stigmata : the respiratory openings in the pharyngeal wall in Molgula, 139.

Stipes : a division of the maxilla in insects, 13.

Stomach : a division of the digestive tract in which digestion goes on; crayfish or lobster, 36 ; Bugula, 86 ; mussel, 96 ; clam, 110 ; snail, 117 ; squid, 130 ; Molgula, 137 ; starfish, 145 ; sea urchin, 152 ; sea cucumber, 156 ; sea anemone, 175.

Stomach-intestine : a division of the digestive tract in which both digestion and absorption go on; grasshopper, 15 ; caterpillar, 21 ; Nereis, 64 ; earthworm, 71.

Stomach pouch : a diverticulum of the stomach; Bugula, 86 ; squid, 131.

Stone canal : a tube joining the madreporic plate with the ring canal in echinoderms, 146, 153, 157.

Submentum : the basal segment of the labium in insects, 13.

Subneural gland : a glandular body in ascidians, 138.

Subumbrella : the oral surface of a medusa, 167, 173, 175.

Supporting layer : the non-cellular layer between the ectoderm and entoderm in Hydrozoa; Hydra, 160 ; tubularian, 165 ; campanularian, 171.

Swimmeret : an abdominal appendage of a crustacean, a pleopod; crayfish or lobster, 32 ; crab, 43.

Symbiotic : the living together of two dissimilar organisms, each being dependent upon the other, 160.

Systemic heart : the median heart of the squid, 129.

Tactile : relating to the sense of touch.

Tarsus : the foot of an insect or a spider; wasp, 4 ; grasshopper, 12 ; spider, 25.

Telson : the terminal segment of a crustacean, 31.

Tentacle : an elongated tactile organ ; Nereis, 62 ; Bugula, 86 ; oyster, 100 ; snail, 114 ; Molgula, 135 ; sea cucumber, 155 ; Hydra, 159 ; tubularian, 164 ; campanularian, 170 ; Gonionemus, 176 ; sea anemone, 178.

Tergite : the dorsal surface of an abdominal segment in insects, 3, 10.

Tergum : the dorsal surface of a thoracic segment in insects; wasp, 3 ; grasshopper, 10.

Terminal : towards or at the posterior or the distal end.

Test : the tunic of the ascidian, 135 : the rigid shell of the sea urchin, 150.

Testis : the male sexual gland; grasshopper, 17 ; spider, 27 ; crayfish or lobster, 37 ; crab, 44 ; copepod, 55 ; Daphnia, 58 ; earthworm, 71 ;

planarian, 77; tapeworm, 83; mussel, 97; oyster, 102; clam, 110; squid, 132; starfish, 146; Hydra, 161.

Thorax: the body-division of arthropods following the head; wasp, 2; fly, 7; grasshopper, 9; caterpillar, 20; spider, 24; crayfish or lobster, 28; sow-bug, 46; amphipod, 48; Caprella, 50; larval decapods, 51; copepod, 53; Daphnia, 56.

Tibia: the segment of an insect's leg between the femur and the tarsus; wasp, 4; grasshopper, 12; spider, 25.

Tiedemann's vesicles: minute diverticula of the ring canal of the starfish, 146.

Trachea: a respiratory tube; grasshopper, 17; caterpillar, 21; spider, 27.

Trichocyst: a cyst containing a defensive bristle in the ectosarc of Infusoria, 186.

Trivium: the three rays of an echinoderm opposite to the bivium, 143, 151.

Trochanter: the segment of an insect's or a spider's leg, between the coxa and the femur; wasp, 4; grasshopper, 12; spider, 25.

Trochophore: a larval form common to polychætous annelids.

Tunic: the outer cuticular covering of tunicates, 135.

Umbo: the protuberance above the hinge on the shell of a lamellibranch; mussel, 89; oyster, 99; clam, 103.

Ureter: a tube forming the outlet of the kidney; crayfish or lobster, 40; mussel, 95; clam, 109; snail, 116.

Uropod: the sixth swimmeret of the macruran decapod, that which forms the swimming tail, 32.

Uterus: a dilated portion of the oviduct in which the egg or the developing animal is detained; planarian, 77; tapeworm, 83.

Vagina: the terminal division of the female reproductive tract; grasshopper, 16; tapeworm, 83; snail, 120.

Vas deferens: a duct leading from the testis towards the external opening; grasshopper, 17; crayfish or lobster, 37; crab, 44; copepod, 55; earthworm, 69; planarian, 77; tapeworm, 83; snail, 120; squid, 132.

Vas efferens: a duct leading from the testis to the vas deferens; planarian, 77; tapeworm, 83.

Vegetative organs: those organs which have to do with the processes of nutrition, growth, and the expulsion of wastes.

Vein: a vessel which brings blood towards the heart; crayfish or lobster, 38; snail, 116; squid, 120.

Velum: the circular muscular membrane of a medusa, 168, 174, 176.

Ventral : on or towards the underside of an animal.

Ventricle : a chamber of the heart from which blood is sent over the body; mussel, 94 ; clam, 108 ; oyster, 101 ; snail, 115.

Vesicula seminalis : a sperm-sac in the male animal; squid, 132.

Viscera : the organs within the body-cavities.

Visceral mass : the compact group of organs comprising the principal viscera in mollusks; mussel, 90; oyster, 100; clam, 104; snail, 113; squid, 125.

Wing-covers : the first pair of wings of a beetle, the elytra, 5.

Yolk glands : planarian, 77 ; tapeworm, 83.

Zoëa : a larval form of the crab and of certain other crustaceans, 51.

Zoœcium : the outer cuticular covering of a bryozoan, 85.

INDEX